CHEMICAL INVESTIGATIONS
FOR CHANGING TIMES

C. ALTON HASSELL
BAYLOR UNIVERSITY

PAULA MARSHALL
BAYLOR UNIVERSITY

CHEMISTRY
FOR CHANGING TIMES

TWELFTH EDITION

JOHN W. HILL • TERRY W. McCREARY • DORIS K. KOLB

Prentice Hall
New York   Boston   San Francisco
London   Toronto   Sydney   Tokyo   Singapore   Madrid
Mexico City   Munich   Paris   Cape Town   Hong Kong   Montreal

Assistant Editor: Jessica Neumann
Editor in Chief, Chemistry and Geosciences: Nicole Folchetti
Marketing Manager: Scott Dustan
Managing Editor, Chemistry and Geosciences: Gina M. Cheselka
Project Manager, Science: Beth Sweeten
Operations Specialist: Amanda A. Smith
Supplement Cover Manager: Paul Gourhan
Supplement Cover Designer: Tina Krivoshein
Cover Credit: BLOOMimage/Getty Images, Inc.

Printed in the United States of America

10 9 8 7 6 5 4 3 2 1

ISBN-13:   978-0-321-61245-8

ISBN-10:   0-321-61245-0

Prentice Hall
is an imprint of

www.pearsonhighered.com

Preface

TO THE STUDENT

Welcome to the chemistry laboratory. During the time that you spend working in this laboratory, we hope that you develop some sense of the excitement that chemists experience. It can be exciting to make something new such as aspirin, especially if you make something useful from waste materials. An example is a water-purifying agent made from an aluminum can. Some people are excited when they see a solid formed from two liquids. Others are fascinated to be able to detect the presence of lead, proteins, vitamin C, or carbohydrates.

A second idea we hope you acquire is that the exciting world of chemistry is all around you; it is everywhere. Everywhere you look, everything you do has some connection with chemistry. It is a chemistry that you can understand, a chemistry in which you can use experiments to investigate your surroundings.

We hope that you experience chemistry in this class. A greater hope is that the excitement and interest you develop will stay with you. Chemistry will continue to be important in your life because it will supply so many things that affect you. We hope that for the rest of your life you continue to experiment with the chemistry around you.

In this lab manual, we have tried to give sufficient background material on each investigation to give you an understanding of the chemical principles you are using. We have also attempted to provide information on waste and the environment. At the end of each investigation, you will notice proper disposal methods for the materials used in the procedure.

TO THE INSTRUCTOR

This edition of the lab manual has continued the style of the last several editions. The idea is to include enough detail in each investigation so the student can do the investigation without much leadership from an instructor. This has been done to free the instructor to do more teaching and to interact more with the students instead of having to demonstrate each procedure. The manual is intended to be a guide for the student, but in no way is it meant to limit you as an instructor. You are the one who can impart the thrill of a discovery, the excitement of a reaction.

We have tried to caution the students about dangerous situations. We have also tried to use as few dangerous chemicals as possible. Ethylene glycol has been eliminated from one experiment and barium from another because of their toxicity. However, there are places where the proper guidance from you will determine the safety of the student. We have also given instructions for the proper disposal of all materials used in the investigation. We hope this will facilitate your job and at the same time give students an appreciation for the care that must be taken to protect our environment. A pre-lab sheet of questions has been added. The intent was to encourage the student to read the experiment carefully before lab. The pre-lab questions are on a separate tear sheet so that it can be turned in as students arrive at lab.

Each experiment and the instructor's manual list the equipment and chemicals needed. The instructor's manual also lists some suggestions about running the experiments. A sample report sheet has been included with the answers that we think are correct. Sample student data are supplied in some investigations, and the amount of time required for each investigation also is listed. A list of recommended equipment for each pair of students follows on the next page.

Basic Equipment List (per pair of students)

glass-stirring rod with policeman
1 suction apparatus (Buchner funnel and filter flask)
beakers: 2 50 mL
         2 150 mL
         1 250 mL
         1 600 mL
         1 1000 mL
graduated cylinders: 1 10 mL
         1 100 mL
Erlenmeyer flasks: 1 125 mL
         2 250 mL
         1 500 mL
test tubes: 6 small (about 10 cm long)
         4 large (about 15 cm long)
         2 extra large (about 25 mm x 200 mm)
stoppers: 1 hole to fit extra-large test tube
         1 hole to fit large test tube
         solid to fit small test tube
         solid to fit large test tube
         2 hole to fit 500 mL Erlenmeyer flask
         1 hole to fit 10 mL graduated cylinder
         1 hole to fit 250 mL Erlenmeyer flask
         1 hole to fit 100 mL graduated cylinder

flint striker or matches
Bunsen burner
2 clamp holders
2 clamps
assorted rubber hoses
glass tubes for holed stoppers
forceps
tongs
test-tube holder
test-tube brush
2 squeeze (wash) bottles
powdered cleanser
spatula
eyedropper
thermometer
evaporating dish
2 watch glasses
2 glass funnels
ringstand
ring
wire and "asbestos" pad

ACKNOWLEDGMENTS

Wow, this is the twelfth edition of this book. One of the most striking thoughts is how many people have been involved with all of those editions. It would be literally impossible to list all those who have contributed ideas and suggestions for this book, yet we wish to recognize a few who have made major contributions by sharing ideas and materials.

The greatest recognition should go to John Hill. He has led us all in how to teach these students. We extend special recognition to Lawrence W. Scott, who led the work through earlier editions. Emerson Garver helped rework the sixth edition. Typist Myriam Roberson is also appreciated. Paul Poskozim and Lynne Cannon provided wonderful in-depth reviews for the eighth edition, Kerri Scott for the eighth and eleventh editions.

Our very patient spouses, Patricia and Les, and our families need special praise for putting up with us during this time. A special thanks again goes to Joshua who helped with some of the figures, and Megan for help in editing and typing.

Most of all, we thank our students at Baylor University, University of Wisconsin-River Falls, Murray State University, and Vanguard College Preparatory School who have class-tested these materials. Their enthusiasm is a special reward to us. May they forever find joy in learning.

C. Alton Hassell
Paula Marshall
Waco, Texas

Contents

Common Safety Regulations

Under OSHA Hazard Communication regulations, a specific set of safety rules must be developed and communicated clearly to all employees. The Department of Chemistry has developed safety regulations that comply with the university's Hazard Communication Program. **Although non-employee students are not covered by these regulations, each student should attest by signature that he or she has read and understands the safety rules.** These rules must be rigorously and impartially enforced. Willful noncompliance should result in dismissal or suspension from the laboratory.

An excellent reference book published by the American Chemical Society is called *Safety in Academic Chemistry Laboratories*. This publication can be obtained by calling the American Chemical Society Referral Service at 1-800-227-5558, option 6. Another source of information is the Chemical Manufacturers Association Referral Center at 1-800-262-8200 (open 24 hours).

General Safety Regulations

The chemistry department strives to provide safe working conditions in all its laboratories. The responsibility for safe working conditions rests with **everyone**, including students, faculty, teaching assistants, and staff. In order to do your part in maintaining safe working conditions, you must obey the General Safety Regulations listed below and report any violations of these regulations that you observe. Such reports should be made to the teaching assistant or to the professor in charge of the laboratory.

1. Eye protection is required at all times in the laboratory and where chemicals are stored and handled. For most laboratory work, the American Chemical Society has suggested that safety goggles are the standard, with safety showers or eyewash fountains near at hand.

2. Long hair and loose clothing must be contained while in the laboratory. Sturdy closed-toe shoes that cover the entire foot, and clothing that covers the body, constitute proper laboratory attire. You will not be permitted to work in the laboratory if you are barefoot or wearing sandals, including sandals with socks. You will also not be permitted in the laboratory if you are wearing shorts, short skirts (minishorts), halter tops, or tank tops. You may wear a lab coat or apron.

3. Work only with materials **after** you have learned about their flammability, reactivity, corrosiveness, and toxicity. This information is readily available through your laboratory instructor and Material Safety Data Sheets. Read your laboratory instructions carefully **before** beginning any laboratory experiment. When in doubt about any part of the procedure, consult your laboratory instructor before initiating the experiment.

4. Unauthorized experiments, including unapproved variations in experiments and changes in quantities of reagents, are forbidden.

5. Never use mouth suction to fill pipets, to start siphons, or for any other purpose.

6. Do not force rubber stoppers onto glass tubing. (Protect your hands with a towel when inserting tubing into stoppers and use water or glycerin as a lubricant.)

7. Wash hands with soap or detergent on leaving the laboratory.

8. Smoking in any part of the laboratory is absolutely forbidden and in many states is against the law.

9. Do not leave experiments in progress unattended without the consent of the laboratory instructor.

10. Horseplay, pranks, or other acts of mischief are especially dangerous in the laboratories and are absolutely forbidden. Sitting on bench tops is not allowed.

11. No eating or drinking is permitted in the laboratory.

12. Students are not permitted to work in the laboratory **unless a teaching assistant is present**.

13. All accidents, no matter how trivial, must be reported. If you have an accident or see an accident, you must report it to the teaching assistant immediately. Never allow an injured person to leave the laboratory to seek assistance without notifying the teaching assistant. A formal written report of all accidents must be made to the Department Safety Officer. The report will include any injuries or damage incurred, the cause, the effect, and any recommendations for prevention of recurrence.

14. In case of an accident, treat all open wounds as a potential source of blood-borne pathogens.

15. If an accident appears to be beyond your control—for example, a large chemical spill or a fire—actuate the fire alarm and vacate the area. If possible, call the University Department of Public Safety and give them the details of the problems—that is, the location and any chemicals involved.

16. Visitors to the laboratories must obey the same safety regulations as the laboratory workers. **This includes the wearing of safety glasses**.

17. Chemical laboratories and storage areas constitute a special hazard for small children. For this reason, **the department does not permit small children in these areas under any circumstances**. Children are not to be allowed elsewhere in the chemistry department except under close adult supervision. The responsibility for the safety of such children must be assumed by the supervising adult. **Under no circumstances should children be left unattended, even for short periods of time, nor should they be brought into areas of risk**.

Chemical Labeling
1. All chemical labels must contain the following information:

- chemical identity
- appropriate hazard warnings
- responsible company and address

2. OSHA standard labels must be used. These labels are available in the stockroom.

3. Proper labels must be on or attached to each hazardous material container.

4. Immediate-use, small-transfer containers are exempt from these regulations.

5. No labels may be removed or defaced without replacements.

6. Labels must be legible, prominent, and in English.

The National Fire Protection Association (NFPA) hazard identification coding system is used in the chemistry department to readily identify chemical hazards. A copy of this labeling system follows.

NFPA Hazard Code Ratings

The NFPA (National Fire Protection Association) ratings provide information relating to health danger (blue), flammability (red), and reactivity (yellow) under emergency conditions such as spills, leaks, and fires. Each section of the label displays a numerical severity rating from 4 (most severe) to 0 (least severe). The fourth section, as stipulated by NFPA, is left blank (white) and is reserved for indicating any unusual reactivity with water or oxidizing properties. The numerical ratings and their associated descriptions are given on the next page.

HEALTH
(Blue)

4 Material that on very short exposure could cause death or major residual injury.

3 Material that on short exposure could cause serious temporary or residual injury.

2 Material that on intense or continued but chronic exposure could cause temporary incapacitation or possible residual injury.

1 Material that on exposure would cause irritation but only minor residual injury.

0 Material that on exposure under fire conditions would offer no hazard beyond that of ordinary combustible material.

FLAMMABILITY
(Red)

4 Material that will rapidly or completely vaporize at atmospheric pressure and normal ambient temperature, or is readily dispersed in air and will burn readily.

3 A liquid or solid that can be ignited under almost all ambient temperature conditions.

2 Material that must be moderately heated or exposed to relatively high ambient temperatures before ignition can occur.

1 Material that must be preheated before ignition can occur.

0 Material that will not burn.

REACTIVITY
(Yellow)

4 Material that in itself is readily capable of detonation or of explosive decomposition or reaction at normal temperatures and pressures.

3 Material that in itself is capable of detonation or explosive reaction but requires a strong initiating source; or that must be heated under confinement before initiation; or may react explosively with water.

2 Material that is unstable and readily undergoes violent chemical change at elevated temperatures and pressures or that reacts violently with water or may form explosive mixtures with water.

1 Material that in itself is normally stable but can become unstable at elevated temperatures and pressures.

0 Material that in itself is normally stable, even under fire-exposure conditions, and is not reactive with water.

SPECIAL WARNINGS
(White)

OX Oxidizing material

W Material that is hazardous when in contact with moisture or water.

Signature Page

Name _____ Class _____ Section _____

I have read and do understand the Common Safety Regulations. I understand that these regulations must be observed for my safety and the safety of other students. I understand that I will not be allowed to work in the laboratory unless I obey these regulations. I agree to abide by these regulations and to report any flagrant violations of these regulations that I observe to the proper authority.

Signature _____

Date _____

1 *Alchemy*

Copper to Silver to Gold

OBJECTIVES

To gain an appreciation for the historical context of some early chemistry.
To produce and observe a chemical reaction resulting in the synthesis of a new substance.
To understand the concept of an alloy.
To understand the concept of density and its use as an analytical tool.
To verify the production of a new substance through measurements and density calculations.

Relates to Chapter 1 of Chemistry for Changing Times, *twelfth ed.*

BACKGROUND

Many of the chemical substances that are currently known, and some of the processes used in chemistry today, were discovered by alchemists. Alchemists discovered alcohol, hydrogen, phosphorus, and gunpowder, as well as the processes of distillation, evaporation, and filtration.

Alchemy was a mixture of science, medicine, trickery or magic, and religion. One of the main goals of alchemy was to change a lesser metal into gold. Producing gold was thought by the alchemists to be a major step toward achieving everlasting life.

Now let's take an imaginary trip back to the Middle Ages when knights were in fashion, there were castles to storm, fair ladies to rescue, dragons to slay (well, maybe there weren't any dragons), and alchemists were at work. The king has just called you for advice. The local alchemist has devised a way to convert copper into silver and then into gold. Your instructions are to perform the experiment, test the gold, and give the king your advice. Does he reward the alchemist or hang him as a cheat?

The alchemist's equations are:

The following table may be helpful for translating the equations into modern English:

| Copper (Cu) | Gold (Au) | Silver (Ag) |

| Zinc (Zn) | Heat or Fire |

The zinc mixed with hot sodium hydroxide becomes a reactive zinc hydroxide complex (a zincate anion $[Zn(OH)_4]^{2-}$) that will adhere to the surface of the copper token as a coating of silver-colored zinc. If there is an impurity on the copper surface, the coating may not adhere properly. Oil from hands and grime from everyday use are examples of impurities that will result in a less-than-perfect product. Both of these can be cleaned easily and quickly from the token using nitric acid. The nitric acid will react only slightly with the copper if the immersion time is short.

When the silver-colored token is heated, the atoms of zinc and copper are excited and will vibrate within their individual positions in the metal lattice. This allows the zinc atoms to migrate into the copper metal forming an intimate mixture, or *alloy*, known as brass. Brass is gold-colored and is used extensively to decorate homes and offices where gold would be too expensive.

Density is defined as the ratio of mass to volume. Each metal has a particular density. A metal object will often be identified by a density measurement. The mass can easily be found with a balance. The volume can usually be calculated if it is a regular-shaped object such as a box ($V =$ length $\times$ width $\times$ depth) or a sphere ($V = \frac{4}{3} \pi r^3$) or a cylinder ($V = \pi r^2 h$). A token is actually a short cylinder. For objects of irregular shape, volume can be found using a liquid displacement method.

Densities:	Copper	9.0 g/cm^3
	Silver	10.5 g/cm^3
	Gold	19.3 g/cm^3
	Zinc	7.1 g/cm^3

Now it's up to you. Remember that the king is very gracious in his rewards for good work, and harsh in his punishments for wrong answers.

WASTE AND THE ENVIRONMENT

Concentrated acids or bases can damage plumbing if they are not diluted and/or neutralized before disposal. Although zinc is not classified as being very toxic, it is better not to introduce metallic ions into the water system. It is safer to dispose of metal ions once they have been tied up in insoluble (solid) compounds by burying them in a secure landfill. The zinc solution should be placed on a metal pan and be allowed to evaporate leaving dry zinc oxide. The oxide can then be placed in a waste container to be buried. **If the zinc solution is placed in the trash, it may cause a fire.** (Wet zinc is a known source of laboratory fires.)

⚠ *CAUTION*s warn about safety hazards.
*EXTRA*s give helpful hints, additional information, or interesting facts.

MATERIALS
Reagents
    zinc, powdered   6 M nitric acid [HNO_3]
    6 M sodium hydroxide [NaOH]
Common Materials
    copper token, wire, or piece of plate   rulers (cm)
    spray can of clear acrylic coating (optional)
Laboratory Equipment
    balance   wire gauze
    caliper   evaporating dish
    ring stand   forceps
    ring   laboratory burner
    stirring rod

2

PROCEDURE

1. Place a small spoonful of powdered zinc (Zn) into an evaporating dish. Cover the zinc with 6 M sodium hydroxide (NaOH) and heat until it begins to steam.

2. Notice the feel of a copper (Cu) token. Measure the diameter of the token and its mass. Record these on the report sheet.

3. Clean the copper token by dipping it in 6 M nitric acid (HNO$_3$) for less than a second. Rinse the token thoroughly with water. Dry the token.

4. Immerse the clean copper token in the steaming alkali solution by sliding it down the side of the dish. When you see bubbles of flammable hydrogen (H$_2$) gas escaping, stop heating. The token will change to a silver color in a short time.

5. Remove the token by pushing it up the side of the dish with a glass stirring rod and grasping it with the forceps. Drop the token into water to cool it and rinse off any remaining zinc powder. Dry and weigh it. Show the token to your instructor so that he or she can initial your report sheet. Record the mass of the token.

6. Grasp the token with forceps and heat gently in the flame of a Bunsen burner. Move the token into and out of the outer cone of the flame. The token may be heated on a hot plate instead. The change from silver to gold will be quite sudden. Heat for only 1 or 2 seconds after the change. Cool the hot token by immersing it in water.

7. After the token cools, dry it, weigh it on the same balance, and show it to your instructor to obtain his or her initials on your report sheet. Record the mass.

8. If so desired, the token can be preserved in either the silver or the gold stage by spraying it with a coat of clear acrylic varnish or by painting it with clear nail polish.

9. Calculate the average thickness of the token by using the weight of the coin and the density of copper as directed on the report sheet. Using the average thickness, calculate the estimated mass of a silver coin and a gold coin.

10. Place the remaining zinc solution in a metal pan and allow the solution to evaporate. Place the dry zinc oxide that forms in a container to be buried in a waste landfill. Flush the used nitric acid down the drain with plenty of water.

⚠️ *CAUTION*
Hot NaOH is corrosive to the skin.

⚠️ *CAUTION*
Nitric acid is corrosive.

EXTRA
A sodium hydroxide solution is basic and is called an alkali solution. The solution may be used for several tokens.

⚠️ *CAUTION*
Don't burn yourself.

EXTRA
Copper can be melted.

EXTRA
This should be done outside the laboratory away from open flames.

⚠️ *CAUTION*
Placing the zinc solution in the trash may cause a fire.

3

ALCHEMY
PRE-LAB QUESTIONS

Name: _____

Lab Partner: _____

Section: _____   Date: _____

1) Legend holds that Archimedes was responsible for proving that a metallurgist had "cut" or diluted the gold for a Ruler's crown with a lesser metal. Archimedes asked the Ruler for a mass of gold identical to what he had provided to the metallurgist. What measurements and calculations did Archimedes most likely make?

2) The alloy produced in this investigation is brass. There are several different types of brass. Use an internet encyclopedia resource to find at least three different types of brass. Describe the differences in proportion and variety of metals used and the effect on the resulting properties of the brass.

3) Consider two different forms of brass. Brass A is 65% copper and 35% zinc and Brass B is 75% copper and 25% zinc. Which form will be the densest? Explain your reasoning.

4) If the crown fit comfortably, what would be the difference between wearing a brass crown and wearing a gold crown of identical mass?

ALCHEMY
REPORT SHEET

Name: _____

Lab Partner: _____

Section: _____   Date: _____

I. MASS OF ORIGINAL TOKEN   _____ g

II. MASS OF DRY SILVER TOKEN   _____ g

Instructor's initials for silver token: _____

How does the silver token feel? (Is it smoother than the copper token? Are there rough spots?)

Is the surface evenly coated with the zinc or are spots of copper still visible?

III. MASS OF COOLED GOLD TOKEN   _____ g

Instructor's initials for gold token: _____

How does the gold token feel?

IV. DECISION
Do you tell the king the token has really turned to silver and then to gold or do you tell him the silver-colored token is just zinc-coated copper and that the zinc has diffused into the copper to make brass, a solid solution of copper and zinc?

EXTRA
Don't lose your head because of a wrong decision.
Consider the densities.

☐ Copper to silver to gold

☐ Copper to zinc coating to brass

V. DENSITY
If you claim the token is not gold, can you prove it by a density calculation?

EXTRA
Density = mass/volume.

Many tokens are not flat, but an average thickness can be calculated from the measured mass and the radius using the equation: $t = m/(\pi r^2 d)$, where r is radius (half the diameter), π is 3.14, t is thickness, d is density, and m is the measured (actual) mass.

Measurements of copper token:

Diameter _____ cm

Radius _____ cm

Mass _____ g

Calculated average thickness _____ cm

You can calculate the volume of a token using the equation volume $(v) = \pi r^2 t$.

Volume of token _____ cm^3

You can calculate the mass of the token using the equation: mass (m) = density (d) × volume (v). The token thickness did not change significantly as it changed from copper to silver to gold, so you can calculate the expected masses of a solid silver and a solid gold token using the volume of the copper token and the densities of silver and of gold. You can then compare those masses with the actual masses.

	Calculated Mass	Actual Mass
Original token		_____
Silver token	_____	_____
Gold token	_____	_____

VI. QUESTIONS

1. If the zinc adhered (stuck to) to the copper instead of being bonded (chemically joined) to it, would the change to silver color be a chemical change?

2. If the heating causes the zinc to bond with the copper, is the change to gold color a chemical change?

3. What happened to the mass of a penny when the U.S. Mint changed the penny's composition from pure copper to a copper-clad zinc coin? (<u>Hint</u>: Consider the densities given in the background material.)

4. What happened to the mass of a dime when the U.S. Mint changed the dime's composition from pure silver to a silver-copper-silver sandwich?

5. How long would it take a person who received an object "changed into gold" to realize they had been the victim of an early magic trick? What would most likely be the first indication?

2 *Density Layers*

Stack the Liquids

OBJECTIVES

To create density gradients and to observe the interfaces between materials.

To observe the positions taken by solid objects placed in a density gradient and then to estimate each object's density from its position.

To calculate the densities of several common household fluids.

To understand the concepts of specific gravity and miscibility.

To consider techniques for separating substances of different phases, miscibility, and densities.

Relates to Chapters 1 and 4 of Chemistry for Changing Times, *twelfth ed.*

BACKGROUND

The *density* of a material is the mass of the material divided by the volume of the material. Density can be expressed in different units depending on the situation. Densities of gases are often stated in grams per liter while the densities of most liquids are given in grams per milliliter.

Because most fluids expand as they are warmed, density depends on temperature, and therefore the specific gravity of a substance is often reported instead of density. *Specific gravity* is the density of the substance divided by the density of water. Since the density of water is very close to 1 g/mL at the usual temperature ranges in which humans live, the value of specific gravity for a particular substance is usually the same as its density, but without units.

A less dense liquid will float on a more dense liquid. This is why petroleum floats on seawater. Liquids of different densities will form separate layers unless they are very soluble in each other. An input of energy, such as stirring, is often required to cause the different molecules to fit among one another — that is, to *dissolve* in each other. Even solutions of different concentrations of the same compound may initially form separate layers that slowly mix to form one homogeneous solution.

Some liquids are not *miscible*; that is to say that they do not dissolve in each other. The amount of miscibility depends upon the similarity of the molecules. Molecules that are alike attract each other and will mix, dissolving in each other. Alcohol and water are miscible with each other but usually *immiscible* with oily liquids. Oily materials are often long-chain organic molecules. Petroleum consists mostly of these long chains and does not dissolve in water. Another example of immiscibility is vinegar-and-oil salad dressing that can be mixed well by shaking, but will separate into distinct layers when left undisturbed.

A solid object will float on a liquid if the density of the object is less than the density of the liquid. Ocean liners are metal but will float on water because the volume of the liner is large enough to make its overall density less than that of water. The inside of the ship is mostly air, which is much less dense than water. A well-known brand of bar soap, Ivory®, has enough air whipped into it to make it less dense than water, so it floats in the bath.

A solid object in a series of liquid layers will float within a layer that is of the same density or will float between two layers with densities higher and lower than its own density.

A material will have different densities in its different *phases*, or states, of matter. Most solids are denser than their liquids. Water is a notable exception in that ice is less dense than liquid water and will therefore float in it.

Often it is necessary to separate substances from one another. Substances of different phase are easily separated from one another using filters, decantation, and skimming. Liquids that are immiscible can be separated by containers that allow selective pouring, much like the gravy cup from your home kitchen with the spout that comes from the bottom of the cup or a cup that has a partial barrier over the spout. Miscible liquids that are in separate layers must be separated without disturbing the *interface*, or area of contact, between the layers. This can be done using specialized equipment that drains from the lower end, such as a buret or a *separatory* funnel.

WASTE AND THE ENVIRONMENT

Most automobile fluids are oily and can contaminate water if not disposed of properly. Ethylene glycol is an alcohol and therefore, poisonous. These fluids should not be poured down the drain or emptied into the environment. Fortunately, service stations are part of an oil recycling program and will accept these fluids and dispose of them properly. Solutions of the household products used in this investigation can be flushed down the drain with plenty of water.

⚠*CAUTION*s warn about safety hazards.
*EXTRA*s give helpful hints, additional information, or interesting facts.

MATERIALS
Reagents
     isopropyl alcohol
Common Materials

antifreeze	brake fluid
30 wt. motor oil	power steering fluid
transmission fluid	mineral oil
Ajax® laundry detergent (liquid)	Downy® fabric softener
Karo® syrup (both kinds: dark and clear)	molasses
vegetable oil	small cork
paraffin	thumb tack
plastic paper clip	aluminum paper clip
rubber stopper or pieces of rubber band	food coloring – green
ice cubes	

Laboratory Equipment

10 mL graduated cylinder	funnel
50 mL graduated cylinder	small test tube
100 mL graduated cylinder	250 mL beaker

PROCEDURE

1. Pour slowly and gently down the side of a small test tube 1 mL each of antifreeze, brake fluid, 30 wt motor oil, power steering fluid, and transmission fluid. Watch as you pour the fluids to see which one goes to which level. Set the test tube aside and allow the liquids to separate.

2. Place 50 mL of dark Karo® into a 250 mL beaker. Tilt the beaker to the side and slowly add 100 mL of vegetable oil so that the oil and syrup do not mix. Gently place a cube of ice on the surface of the top liquid. Note in the Report Sheet the position of the ice. Set the beaker aside until after Procedure 8. Note the changes in the beaker on the Report Sheet

3. Weigh a clean, dry 10 mL graduated cylinder. Fill the 10 mL cylinder with about 10 mL of mineral oil. Read the volume to two decimal places. Measure and record the mass of the cylinder and oil.

4. Pour the mineral oil into a 100 mL cylinder.

5. After cleaning and drying the 10 mL cylinder, measure and record the volume and weight of about 10 mL of isopropyl alcohol. Add several drops of green food color to the alcohol.

6. Using a long funnel or tubing that touches the bottom of the 100 mL cylinder, pour the isopropyl alcohol gently underneath the mineral oil.

7. Repeat steps 5 and 6 using, in order, Ajax® laundry detergent, Downy® fabric softener, clear Karo® corn syrup, and molasses. You do not need to add food color. Continuing with the funnel in place, gently place each new liquid on the bottom of the cylinder. As the liquids become denser, more time will be required for them to drain out of the funnel. After the molasses has drained out, remove the funnel in a swift pull.

8. Determine the approximate density of a thumb tack, an aluminum paper clip, a small rubber stopper or a piece of rubber band, a plastic paper clip, a piece of paraffin, and a cork by allowing the objects to slide gently down the side of the cylinder. Each object will stop at the level of its density or between liquids that are higher and lower in density.

9. Clean the inside of the top of the auto fluids test tube prepared in step 1 with a paper towel. Snugly insert a rubber stopper. Gently invert the test tube and set it into a beaker. Observe which fluids mix and which form new layers.

10. Pour the automobile fluids into a marked waste container. You may stir together the liquids in the 100 mL cylinder and pour them down the drain.

EXTRA
The less mixing of the liquids, the better the investigation.

EXTRA
The funnel can be left in place until all liquids have been added.

EXTRA
Sliding the objects down the side decreases the disturbance of the solution layers.

EXTRA
The auto waste should be recycled at a service station.

9 Explain d
mL, less, more

DENSITY LAYERS
PRE-LAB QUESTIONS

Name: _____

Lab Partner: _____

Section: _____   Date: _____

1) When you are walking through a parking lot and see a puddle of water with a rainbow effect on the surface, you are observing a thin-film optical phenomenon. Do you suspect the material in the film is miscible or immiscible with the water? Explain your reasoning.

2) What is the nature of film on the surface of the water in the first question and what was its likely source?   *oil, Gas*

3) Assume you had to separate a series of layers of materials that have different densities. Suggest a possible method for separating each of the following:
   a) solid objects of greater density than the liquid in which they are contained   *Decant liq off*

   b) solid objects of less density than the liquid in which they are contained   *Skim off top*

   c) two immiscible liquids of significantly different densities   *Decant*

   d) two miscible liquids of significantly different densities   *Settle – decant*

4) We should be careful not to allow petroleum products to get into our waterways, and yet each time it rains we see evidence of those products in the environment. Can you suggest a creative method to trap these products as they run into street sewers?   *Carbon/Charcoal Absorbers*

5) Why is it not a problem for vegetable oil to enter waterways when motor oils are an environmental hazard?   *Bio degradable*

DENSITY LAYERS
REPORT SHEET

Name: _____

Lab Partner: _____

Section: _____   Date: _____

I.   AUTOMOBILE FLUIDS
List the fluids in order of decreasing density (most dense first).

_____

_____

_____

_____

_____   Instructor's initials: _____

II. ICE ON LIQUIDS
At what position is the ice originally?

After Procedure 8, where is the liquid water that came from the melted ice?

III. HOUSEHOLD LIQUIDS
Mass of 10 mL graduated cylinder   _____ g

Mineral oil
Mass of cylinder and oil   _____ g   Volume   _____ mL
minus mass of cylinder   − _____ g   _____ g
                                               Density   _____ = _____ g/mL
Mass of oil   _____ g   _____ mL

Isopropyl alcohol
Mass of cylinder and alcohol _____ g   Volume   _____ mL
minus mass of cylinder   − _____ g   _____ g
                                               Density   _____ = _____ g/mL
Mass of alcohol   _____ g   _____ mL

Ajax laundry detergent
Mass of cylinder and Ajax   _____ g   Volume   _____ mL
minus mass of cylinder   − _____ g   _____ g
                                               Density   _____ = _____ g/mL
Mass of Ajax   _____ g   _____ mL

15

Downy fabric softener

Cylinder and Downy	_____ g		Volume	_____	mL
minus cylinder	− _____ g				
		Density	$\dfrac{\text{_____ g}}{\text{_____ mL}}$ =	_____	g/mL
Mass of Downy	_____ g				

Karo

Cylinder and Karo	_____ g		Volume	_____	mL
minus cylinder	− _____ g				
		Density	$\dfrac{\text{_____ g}}{\text{_____ mL}}$ =	_____	g/mL
Mass of Karo	_____ g				

Molasses

Cylinder and molasses	_____ g		Volume	_____	mL
minus cylinder	− _____ g				
		Density	$\dfrac{\text{_____ g}}{\text{_____ mL}}$ =	_____	g/mL
Mass of molasses	_____ g				

Object	Estimated Density		Object	Estimated Density	
Thumb tack	_____	g/mL	Aluminum paper clip	_____	g/mL
Rubber stopper/band	_____	g/mL	Plastic paper clip	_____	g/mL
Paraffin	_____	g/mL	Cork	_____	g/mL

IV. QUESTIONS

1. Why shouldn't the automobile fluids be poured down the drain? *Non-Bio*

2. Of the automobile fluids, which ones mix well with one another? Which fluids form separate layers?

3. After observing the differences in the densities of the auto fluids, explain why it is important that the liquids used in an automobile be uncontaminated. *each a purpose*

4. What does it mean if a solid object sinks to the bottom of a cylinder filled with a liquid? *More*

5. What does it mean if a solid object floats on top of a liquid? *less*

6. Karo and molasses are mostly sugar in water. Why do they form separate layers? *more sugar*

7. Based upon your calculations in part III, which of the household liquids will float on water? *oil*

8. What can be inferred about the relative densities of ice and water from the ice-on-liquids procedure? Give an estimate of the value of the density of ice. *d ice < d H₂O*

3 Energy in Physical and Chemical Changes

Easy Come, Easy Go

OBJECTIVES

To gain an understanding of the basic types of energy changes that occur when atoms rearrange.

To make quantitative observations of positive and negative heats of solution.

To make quantitative observations of positive and negative heats of reaction.

To understand that all changes involve energy, even though that energy may or may not be exchanged with the surroundings.

To make measurements of a series of energy changes to illustrate that energy is conserved.

Relates to Chapters 1, 4, and 15 of Chemistry for Changing Times, *twelfth ed.*

BACKGROUND

Energy is involved any time atoms rearrange. The atoms may be part of a *physical change* such as a change of state or a change of condition, as when a substance dissolves, forming a solution. The atoms might also be involved in a *chemical change* in which a new substance is produced.

Energy is required any time particles separate from one another in moving from a state of lower energy to one of higher energy. For example, when ice melts or when water evaporates, the particles must overcome the energy that holds them together. That energy is part of a system of *intermolecular forces*. Change of state is a form of physical change with which we are all familiar.

Another form of physical change occurs when a substance dissolves in a solvent. Some particles in a substance are held together in a crystalline state by ionic bonds, or *lattice energy*. Table salt is an example of an ionic crystal. Other substances are held together in the liquid or solid state by intermolecular forces such as hydrogen bonding or other electrostatic attractions. Sugar crystals, ice, liquid water, and alcohol are all examples of substances held together by these intermolecular forces. When the particles separate from one another, energy must be absorbed in order for them to overcome these forces and move apart. This absorption of energy is an *endothermic* process.

Energy is released when particles form an attachment of some kind to another particle. This is an *exothermic* process. As water vapor condenses to form liquid water, or as liquid water solidifies to form ice, energy is released. Likewise, as particles of a dissolved substance form attachments to the particles of a solvent, energy is released. If the energy released is more than the energy absorbed when the particles are separated from one another, then the overall, or net, energy change is exothermic and the energy goes to the surroundings. This energy release would be observed as a vessel that felt warm to the touch. If, however, the energy released as the particles are attracted to the solvent is less than the energy required to separate them, then the net energy change is endothermic and the vessel would feel cool to the touch because energy is absorbed from the surroundings.

Chemical changes follow the same idea. When bonds are broken, energy is absorbed. When bonds are formed, energy is released. If the amount of energy absorbed when the bonds of the original substances are broken is greater than the energy released when the bonds of the new substances are formed, the net energy change for the reaction is *endothermic*. If the energy absorbed is less than the energy released, the net energy change is *exothermic*.

In some reactions the energy changes are equal in magnitude but opposite in direction so that no exchange of energy with the environment can be detected. These processes are referred to as being *isothermic*.

We will look at several physical and chemical changes and note the energy changes that accompany them. We will also make measurements of temperature changes for these processes to show that energy is always conserved in physical and chemical processes. Note that our temperature measurements will not be error free as a result of the simplified apparatus we will use. However, the measurements are intended to be used for general observations only.

WASTE AND THE ENVIRONMENT
All solutions from this investigation can be washed down the drain with plenty of water.

▲ *CAUTION*s warn about safety hazards.
*EXTRA*s give helpful hints, additional information, or interesting facts.

Reagents
    ammonium chloride [NH_4Cl] solid
    calcium chloride [$CaCl_2$] solid
    sodium hydroxide [NaOH] solid
    1.0 M sodium hydroxide [NaOH] solution
    sulfuric acid [H_2SO_4] concentrated
    1.0 M hydrochloric acid [HCl]
    0.5 M hydrochloric acid [HCl]

Common Materials
    Styrofoam cups
    cardboard (for lid)
    wire for stirrer

Laboratory Equipment
    weighing paper or weighing boats
    thermometer
    100 mL graduated cylinder

PROCEDURE
Part A: Heats of Solution and Dilution – Physical Changes
HEAT OF SOLUTION FOR AMMONIUM CHLORIDE

1. Assemble a coffee cup calorimeter using a Styrofoam cup, cardboard lid, and thermometer like the one in the diagram to the right. (If the thermometer does not remain firmly at the level where you want it, you can add a rubber stopper that has been split or bored to fit the thermometer so that the stopper rests on the cardboard lid. A rubber band works well, also.)

2. Place 30. mL of distilled or deionized water into the Styrofoam cup and allow it to sit until the temperature stabilizes. Record the temperature.

3. Add 6.0 g of ammonium chloride (NH_4Cl) to the water, quickly assemble the cardboard lid, thermometer, and stirrer into the calorimeter. Stir by moving the stirrer carefully up and down. Periodically check to see if any solid is left. When the solid is dissolved, continue to agitate until the temperature stabilizes. Record the temperature. Clean and dry the Styrofoam cup.

HEAT OF SOLUTION FOR CALCIUM CHLORIDE

4. Repeat steps 1 through 3 using 9.0 g of calcium chloride ($CaCl_2$).

HEAT OF SOLUTION FOR SODIUM HYDROXIDE

5. Repeat steps 1 through 3 using 100. mL of water and 2.0 g of sodium hydroxide (NaOH).

HEAT OF DILUTION FOR SULFURIC ACID

6. Place 30. mL of water into the Styrofoam cup and measure the temperature.

7. Carefully add 5 mL of concentrated sulfuric acid (H_2SO_4) to the water in the cup and quickly assemble the top and thermometer. Stir to thoroughly mix the two solutions, and record the final temperature of the diluted acid solution. Clean and dry the cup.

Part B: Heats of Reaction – Chemical Changes

8. Place 50. mL of 1.0 M HCl (hydrochloric acid) into the cup and record temperature after it has stabilized.

9. Add 50. mL of 1.0 M NaOH and quickly assemble the calorimeter. Stir to completely mix the solutions. Allow the temperature to stabilize, and record it. Clean and dry the cup.

10. Place 100. mL of 0.5 M HCl into the Styrofoam cup and allow the temperature to stabilize. Record the temperature.

EXTRA
In order to prevent heat exchange with the surroundings, place the thermometer and the top on the calorimeter quickly.

⚠ *CAUTION*
Sodium hydroxide is caustic and should not come into contact with your skin. Wear gloves.

⚠ *CAUTION*
Sulfuric acid is especially corrosive to living tissue. Wear gloves.

EXTRA
Always pour acid into water, not water into acid, when making dilutions.

⚠ *CAUTION*
Hydrochloric acid is corrosive to the skin.

11. Add 2.0 g of solid NaOH to the solution and quickly assemble the Styrofoam cup, thermometer, and lid. Stir to completely dissolve and react the solid. When the temperature stabilizes, record the final temperature.

▲ *CAUTION*
Sodium hydroxide is caustic and should not come into contact with your skin. Wear gloves.

ENERGY IN PHYSICAL AND CHEMICAL CHANGES
PRE-LAB QUESTIONS

Name: _____

Lab Partner: _____

Section: _____   Date: _____

1) In a certain ionic compound, more energy is required to overcome the lattice energy holding ions together in the crystal than is released when the ions dissolve in water. Will the overall process appear to be endothermic or exothermic? Explain your logic.

2) Consider two solids that are stirred together in a glass beaker at room temperature and interact to produce a liquid while frost is seen to form on the outside of the beaker. Is the process endothermic or exothermic? Discuss the energy changes that can be assumed in the process.

3) Is the resulting liquid in question #2 most likely the result of a physical or chemical change? Use the magnitude of the energy changes to support your answer.

4) Two substances in solution at room temperature are mixed in a test tube containing a thermometer to register temperature changes. There is no change on the thermometer but the combined solution immediately becomes cloudy and a white powder eventually settles to the bottom of the test tube. Why is it incorrect to determine that no energy changes occurred?

5) Ice and hot water are combined in the same container. Discuss the energy changes that will take place and any state changes that will result.

6) If ice remains after the temperature stabilizes in the container from question #5, what can be assumed about the final temperature?

21

ENERGY IN PHYSICAL AND CHEMICAL CHANGES
REPORT SHEET

Name: _____

Lab Partner: _____

Section: _____   Date: _____

I. PART A: Heat of Solution and Dilution —Physical Changes

	TEMPERATURE		
	INITIAL	FINAL	Δt (final – initial)
Ammonium chloride	_____	_____	_____
Calcium chloride	_____	_____	_____
Sodium hydroxide	_____	_____	_____
Sulfuric acid	_____	_____	_____

II. PART B: Heat of Reaction —Chemical Changes

	TEMPERATURE		
	INITIAL	FINAL	Δt (final – initial)
HCl and NaOH solution	_____	_____	_____
HCl and solid NaOH	_____	_____	_____

III. QUESTIONS
1. Is there a difference in the final solution temperature of the HCl with NaOH solution and the HCl with solid NaOH? Can the difference be explained by the heat of solution of NaOH? (Hint – Compare the temperature changes for the HCl with solid NaOH to changes for both the water with solid NaOH and the HCl with NaOH solution.)

2. Explain why acid is poured into water instead of water into acid, given the fact that if drops splatter during pouring, the drops are usually from the liquid in the container, not the liquid being added.

3. Could a solution of one of these compounds serve as a cold pack for an athletic injury? If so, explain.

4 *Diffusion and Graham's Law*

It's Not About Crackers

OBJECTIVES

To observe the relative rates of diffusion of particles in a liquid at different temperatures.

To determine the relative rates of diffusion of two gases at room temperature and normal atmospheric pressure.

To compare the experimental rate ratio to the expected ratio from Graham's law.

To understand the factors that affect a diffusion rate.

Relates to Chapter 2 and 5 of Chemistry for Changing Times, *twelfth ed.*

BACKGROUND

One of the best pieces of evidence for the particulate nature of matter and support for the kinetic molecular theory is the process of *diffusion*. When you wake up in the morning and know that someone has put on the coffee pot, or that there will be bacon for breakfast, you are detecting gaseous molecules that have diffused through the atmosphere from their source to where you are when you detect them. This movement is a result of the constant random motion of the particles. Diffusion occurs in gases, in liquids, and even in solids. Particles in a solid *vibrate* but remain in their position for the most part. This vibration allows them to occasionally swap places with a neighboring particle. In liquids, the particles vibrate, but they also *rotate*, and this second motion makes it easier for them to slip past one another. In gases, however, the particles vibrate, rotate, and *translate*, or move from one location to another. It is this third type of motion that encourages diffusion to take place on a larger scale in gases.

Since the motion of particles in the gaseous phase is less restricted than the motion of particles in the liquid or solid phase, the rate of diffusion in a gas is much greater than in the other two phases. Because of the relatively great distances between the particles compared to their sizes, they have very little effect on one another and are therefore free to move about quickly. Gas particles move at tremendous velocities depending upon their *temperature*. Temperature is our way of comparing the average kinetic energies of particles; hence, the higher the temperature, the faster the particles are moving. The oxygen molecules in the room right now, assuming the temperature to be somewhere near 25°C, are moving at an average of more than 600 meters per second or 1900 feet per second.

Particles do not move across a space at that high rate of speed, because they are moving in erratic, random motions caused by collisions with other molecules and with obstacles in that space. Therefore, as gas particles move from their point of origin, they fan out in all directions and slowly make their way across the space. If you were able to follow the path of one particle, you would find it to be very erratic. The particle would move in one direction for a distance only until it encountered another particle. The resulting collision would send it in another direction until it collided with another particle, or wall, or table, or your skin. As a result, the rate at which a particle makes its way in one particular direction can be quite slow.

Additionally, these collisions are the source of the gas property we call *pressure*. The number of collisions with a surface in a period of time, together with the average speed of the particles at collision, determines the total pressure exerted on a surface. Therefore, both temperature and the number of particles in a container determine pressure. In addition, pressure can help determine the rate of diffusion of a gas.

A third factor that determines the rate of diffusion in a gas is the *mass* of the individual particles. Because particles at the same temperature have a common average kinetic energy, the more massive particles must be moving at a slower rate. (Kinetic energy is the product of one-half the particle's mass and the square of its velocity.)

In this investigation, we will look at the relative rates of diffusion of particles in the gaseous and liquid phases and examine the effects of temperature and particle mass on the rates.

WASTE AND THE ENVIRONMENT
All final substances in this investigation are nontoxic and can be washed down the drain with plenty of water.

▲ *CAUTION*s warn about safety hazards.
*EXTRA*s give helpful hints, additional information, or interesting facts.

Reagents

hydrochloric acid [HCl] concentrated	distilled water
ammonia [NH_3] concentrated	

Common Materials

cotton swabs	ice cubes
food coloring (blue or green)	ruler (cm)

Laboratory Equipment

1" glass tube	4 250 mL beakers
2 one-hole stoppers to fit tube	thermometer
stopwatch	2 watch glasses or evaporating dishes
grease pencil	barometer

PROCEDURE

Part A: The Effect of Temperature on Diffusion in Liquids

1. Obtain three 250 mL beakers. Fill one with water that is at room temperature. Add a couple of ice cubes to the second beaker filled with water. Warm the third beaker of water to 60°C.

2. When the third beaker has reached the desired temperature, remove any remaining ice from the second beaker. Assume the water in the second beaker to be at 0°C.

3. Place one drop of green or blue food color into the top of each beaker of water. Record the time on the report sheet.

4. Observe the relative rates of diffusion as the color migrates through the water in each beaker. When the color is uniformly distributed through the water in a beaker, record the time.

5. Occasionally check the progress of the three beakers as you begin the next section of the investigation.

EXTRA
Any color will do, but the darker colors are easier to distinguish as they are diluted. If you choose to use red or yellow, use two or three drops in each beaker.

Part B: The Effect of Particle Mass on Diffusion in Gases

6. Obtain a 1-in. diameter glass tube and two one-hole stoppers to fit the ends of the tube.

7. Cut four or five cotton swabs in half and place them on a clean paper towel.

8. Obtain two watch glasses or evaporating dishes in which to carry the swabs once they have been moistened.

9. Fill a 250 mL beaker with water for the purpose of neutralizing the cotton swabs after each trial.

10. Moisten one cotton swab with concentrated hydrochloric acid (HCl) and one with concentrated ammonia solution (NH_3) by dipping the swabs into the containers provided by your instructor.

11. Place each of the cotton swabs into the narrow end of a stopper with the moistened tip toward the tube ends. Simultaneously place each stopper into the ends of the tube. Begin timing with a stopwatch. Be sure the tube remains motionless while the diffusion is in process.

12. As soon as you detect a white smoke ring in the tube, stop the stopwatch, mark its position with a grease pencil, and also mark the position of the two tips of the cotton swabs on the glass. The ring will continue to migrate after it forms, so mark it quickly. The white ring is ammonium chloride (NH_4Cl).

⚠ *CAUTION*
Do step 10 immediately before you are ready to place the swabs into the ends of the tube (step 11), as we want to allow very little of the gas to evaporate into the room. Both substances are caustic and very irritating to the internal membranes if inhaled.

13. IMMEDIATELY remove the cotton swabs into the beaker of water. Rinse the tube quickly.

14. Measure the distance in cm between the cotton swab and the position of the ammonium chloride ring for each gas. Record the distances ("D") and the time in seconds ("t") in the data tables provided.

15. Dry the interior of the tube with a long, twisted paper towel. Clean the grease marks from the exterior of the tube with a paper towel.

16. Repeat the experiment three more times.

17. Record the current barometric pressure and temperature on the data page.

18. Clean the glass tube thoroughly, removing all marks, and dry it completely. Return all equipment to your instructor.

19. Remove the cotton swabs from the water and place them in the trash can. Pour the water from all four beakers down the drain.

20. For each trial, calculate the rate of diffusion ("R") for each gas and record it in the data table.

21. Calculate the molar mass for each gas and record it in the data table.

22. Calculate the average diffusion rate for each gas and record it in the data table.

23. Calculate the *experimental* rate ratio by dividing both rates by the smallest to express the ratio as "1: ___."

24. Calculate the *expected* rate ratio using Graham's law. A mathematical version of Graham's law is

$$\frac{\text{rate}_{HCl}}{\text{rate}_{NH_3}} = \frac{\sqrt{\text{molar mass } NH_3}}{\sqrt{\text{molar mass } HCl}}.$$

Use "1" as the "rate$_{HCl}$" in the equation and solve for the expected rate of NH_3. Record the expected rate ratio on the data table.

⚠ *CAUTION*
Do not place the cotton swabs directly into the trash!

EXTRA
Moisture will adversely affect the results in subsequent trials.

DIFFUSION AND GRAHAM'S LAW
PRE-LAB QUESTIONS

Name: _____

Lab Partner: _____

Section: _____   Date: _____

1) Give several examples of diffusion from your own experience.

2) If gas molecules are moving at a tremendous speed, why does it require several minutes for a scent to diffuse across a room?

3) How would an increase in atmospheric pressure be expected to affect the rate of diffusion in a gas?

4) Explain how an increase in temperature would be expected to affect the rate of diffusion in a gas.

5) How does the molar mass of gas particles affect the relative rates of diffusion?

DIFFUSION AND GRAHAM'S LAW
REPORT SHEET

Name: _____

Lab Partner: _____

Section: _____   Date: _____

I. DATA AND CALCULATIONS
PART A: The Effect of Temperature on Diffusion in a Liquid

Beginning time: _____

	Beaker #1	Beaker #2	Beaker #3
Temperature			
Ending time			
Elapsed time			

PART B: The Effect of Particle Mass on Diffusion in Gases

	Trial #1	Trial #2	Trial #3	Trial #4
Time, t (in sec)				
D (HCl to ring) (cm)				
D (NH$_3$ to ring) (cm)				
R (HCl) (cm/sec)				
R (NH$_3$) (cm/sec)				

	Molar Mass (g/mol)	Average Rate (cm/sec)
HCl		
NH$_3$		

Rate Ratios
Experimental: _____

Expected: _____

II. QUESTIONS

1. What was the effect of the different temperatures on the rates of diffusion in the beakers of water?

2. Did you notice any "currents," or convection, as the color moved through the water? If so, account for the cause of the currents.

3. Write the balanced equation for the formation of ammonium chloride.

4. Use the experimental ratio and the accepted ratio to calculate percent error.

    Percent error = $\dfrac{|E - A|}{A} \times 100\%$ where "E" is the experimental ratio and "A" is the expected ratio.

5. List possible sources of error that account for the percent error in your experimental data.

5 *Atoms and Light*

Light Is Not Just Light

OBJECTIVES

To observe the colors of light emitted from excited atoms of different elements.

To become familiar with several methods of exciting electrons in atoms.

To understand that the unique spectrum of each element is an efficient and valuable analytical tool for chemists.

To learn to use a diffraction grating to separate the emission spectrum into discrete wavelengths of light.

To relate these wavelengths to the energy changes of the electrons within the atom.

Relates to Chapter 3 of Chemistry for Changing Times, *twelfth ed.*

BACKGROUND

Atoms are made up of positive nuclei surrounded by negative electrons. These electrons have different amounts of energy as a rule, but the important fact is that electrons cannot possess just any amount of energy, but only certain amounts of energy. How do we know that? This knowledge comes from studies of *electromagnetic radiation* in the X-ray, ultraviolet, and visible portions of the spectrum. In this investigation we will make detailed observations of light coming from atoms and these observations will lead us to the same conclusion. We will make observations based upon common light sources such as a candle, incandescent bulb, and fluorescent bulb. We will also observe light emitted from atoms in flame tests and high voltage discharge tubes.

When an atom is heated, subjected to an electric current, or placed in the path of high energy radiation, electrons absorb energy in definite amounts, becoming *excited*. They can reemit that extra energy as electromagnetic radiation. We perceive each emission of energy as a particular color of light. Electrons in different elements absorb and then emit different amounts of energy, producing different spectra. The observed *spectra*, or colored lines, can be used to identify the element. Helium was discovered in the Sun before it was known on Earth by a study of the spectra from the sun. The temperature of a star is indicated by the color of light it emits. The temperature, in turn, determines the mass of heaviest element that star can form.

A *diffraction grating* is a sheet of material with closely spaced grooves that works like a prism in that it separates the different wavelengths of light. Looking through a diffraction grating, white light appears as a complete rainbow, or a *continuous spectrum*. The light coming from an excited atom will appear to be a set of distinctly colored lines. Without a diffraction grating, only the most prominent color is seen. Atoms with many electrons, like tungsten, give off so many colors of light that we observe a complete spectrum — that is, white light. For this reason, tungsten is the element used in the common incandescent light bulb. It is interesting to note that electrons can have wave properties as well as particle properties, just as light does. The properties of physical waves lead to the same conclusion: Electrons around nuclei can have only certain energies.

Instruments such as the *atomic absorption spectrometer* make use of the fact that every element has a particular set of energies that are absorbed when its electrons are excited. These particular energies of light can be used to detect the presence of the particular element. The energy of a particular light ray can be related to a wavelength of light.

Other instruments identify elements by the wavelengths of light that are re-emitted when the atoms are excited. For example, in a *flame test*, a solution containing the atoms of a particular element is heated in a flame, and the spectrum produced can be used to identify the element. The spectrum contains the colors produced by the energized, or *excited*, electrons in the heated element. *Discharge tubes* contain an element in its gaseous state at a very low pressure and run a high voltage electric charge through the tube. The electric energy excites the electrons in the atoms and the electrons in turn re-emit the energy as a particular color of light.

One element whose behavior is slightly different is europium, Eu. When it is bombarded by ultraviolet light, it absorbs a certain color (energy), as the other elements do, but when europium re-emits the energy, it does so in two steps. The energy of one step is in the visible region of light, producing a rose-pink glow. The second step occurs by a radiationless transition. The compounds of europium are used to produce the red color in a color TV tube. The electrons in the compound absorb energy from the ultraviolet light, then re-emit the energy as red light.

WASTE AND THE ENVIRONMENT
Because the solutions for the flame tests and the europium oxide will be kept for next semester's session, only the concentrated acid poses a hazard. Concentrated acid can damage plumbing if not neutralized or diluted.

⚠ *CAUTION*s warn about safety hazards.
*EXTRA*s give helpful hints, additional information, or interesting facts.

Reagents
    magnesium ribbon
    0.1 M chloride salts of Li, Na, Ba, Sr, K, and Cu
    solid europium oxide [Eu_2O_3]

Common Materials
    candle
    incandescent bulb in socket

Laboratory Equipment
    flame test wires
    diffraction gratings
    hydrogen discharge tube
    helium discharge tube
    neon discharge tube
    mercury discharge tube
    ultraviolet light
    high-voltage discharge apparatus
    earth light (optional)

PROCEDURE

(*Note*: The room needs to be darkened for best results in this investigation.)

1. Hold a diffraction grating up to your eye and view the candlelight through it. You may need to look to the left or right through the diffraction grating to see the spectra. Record your observations.

2. Using the diffraction grating, view each of the following light sources when demonstrated by the instructor:
    Burning magnesium strips
    Incandescent bulb
    Hydrogen discharge tube
    Helium discharge tube
    Neon discharge tube
    Mercury discharge tube
    Record your observations for each light source.

3. Without the grating, observe the colors as the instructor places different metal salts into a flame. Record these observations.
    Lithium
    Sodium
    Barium
    Strontium
    Potassium
    Copper

4. Observe and record your observations as the instructor shines ultraviolet light on europium oxide.

▲ *CAUTION*
Do not look directly at the burning magnesium strip.

EXTRA
Flashbulbs use magnesium, and the filament of an incandescent bulb is tungsten.

ATOMS AND LIGHT
PRE-LAB QUESTIONS

Name: _____

Lab Partner: _____

Section: _____   Date: _____

1) Name several ways in which atoms can be excited.

2) When a fuel such as wood is burning in a fireplace, visible light is one of the products of the reaction. What is the source of this light we call the "flame"? (Give your answer in the form of a sequence of events.)

3) How might the normal color of a flame mask the true color of a salt during a flame test?

4) When the electric element on a home cook top is turned on, it first becomes light orange and then turns red hot. If more current is supplied, it can become hotter. If there was no restriction on the amount of current, at some point the temperature would be sufficient for the burner to glow white. What has happened to the number of individual colors of light being emitted if the burner appears white to the naked eye?

5) Video cameras measure colors in Kelvin temperatures. Why is this appropriate?

ATOMS AND LIGHT
REPORT SHEET

Name: _____

Lab Partner: _____

Section: _____   Date: _____

I. Describe the spectrum of each of the following as seen through the diffraction grating:
  1. Candlelight

  2. Burning magnesium

  3. Incandescent bulb

  4. Hydrogen

  5. Helium

  6. Neon

  7. Mercury

II. Describe the color of the flame produced by the salts containing these metals:
   1. Lithium

   2. Sodium

   3. Barium

   4. Strontium

   5. Potassium

   6. Copper

III. QUESTIONS
   1. Are the results the same for candlelight, burning magnesium light, and the incandescent (tungsten) bulb?

   2. Could the colors of the flames be used to identify the different metal ions?

   3. Explain why a yellow shirt appears yellow.

   4. Large mercury vapor lamps are often used as night time security lights. What color would you expect these to appear to the naked eye?

   5. What color would you expect from a sodium vapor lamp such as those used in some parking lots?

   6. What applications can you think of for europium compounds?

   7. What can we learn from a study of light from different stars?

6 *Flame Tests and Analysis*

What's It Made of?

OBJECTIVES

To learn to identify specific nonmetal anions contained in a solution.

To become familiar with the flame-test procedure for metal cations.

To learn to read and interpret a qualitative analysis scheme.

To use a qualitative analysis ("qual") scheme to identify correctly the ions present in an unknown salt.

Relates to Chapter 3 of Chemistry for Changing Times, *twelfth ed.*

BACKGROUND

One of the major fields of chemistry is analytical chemistry. This field is made up of two main areas: quantitative analysis (determining the concentration, or amount, of a substance) and qualitative analysis (determining which elements or compounds are present). This investigation will focus on qualitative analysis as we learn to identify the particular ions present in an unknown salt solution.

Many compounds are ionic—that is, made of anions and cations. Sodium chloride is an example of an ionic substance. *Cations* are positive ions, generally from the metallic components of ionic compounds that have been dissolved in water, such as the sodium ion, Na^+. *Anions* are negative ions present either as monatomic nonmetal ions, such as Cl^-, or as polyatomic ions. *Polyatomic ions* are groups of atoms that are bonded together to form a unit that has a charge. Examples of polyatomic ions are sulfate, SO_4^{2-}, and nitrate, NO_3^-.

The cations in this investigation will be limited to sodium ions, potassium ions, lithium ions, and copper ions. One method of determining which cation is present in a compound is by a flame test. Each ion produces a particular color when it is placed into a flame. The flame excites electrons within the atoms to higher energy levels. When the electrons return to the original lower level, the energy given off is in the form of visible light and is a color unique to that element. The unique yellow-orange color of a sodium vapor lamp is due to this phenomenon.

Two methods for the flame test are described. The first method makes use of a traditional wire tool called a *flame loop*, which holds a thin film of solution. The other method involves soaking a piece of paper in the solution to be tested and holding the paper in a flame. If the paper method is used, the paper should not be allowed to catch fire. As long as it remains wet, the kindling temperature of the paper (the temperature at which it catches fire) will not be reached, and the paper will not burn.

In this investigation the anions will be limited to four: chloride, Cl^-; sulfate, SO_4^{2-}; nitrate, NO_3^-; and acetate, $C_2H_3O_2^-$. The presence of a particular anion can sometimes be determined by reactions unique to the anion. Often the determination involves a scheme, or a series of reactions, the results of which allow anion identification. When the ions of normally insoluble ionic compounds are placed into the same solution, they form a solid called a *precipitate*. A precipitate is first seen as a clouding of the solution by a white or colored powder that will eventually settle to the bottom of the container. Often a precipitation reaction can identify what ion or ions are present in a solution. For example, if a solution contains lead ions, the addition of chloride ions will precipitate the lead as lead chloride crystals. This is because lead chloride is an insoluble salt.

All the unknown salts in this investigation will be soluble in water. We will first make a solution of the salt. In testing the unknown solutions to identify the anions present, part of the original solution can be tested for nitrate by what is called the *brown ring test*. A small portion of the solution is added to an iron sulfate solution. Sulfuric acid is layered below the iron sulfate solution. Any nitrogen in a nitrate ion located at the interface is reduced by the sulfuric acid and forms the complex nitrosyliron (II) ion ($Fe(NO)^{2+}$), which is brown in color. If there is no brown ring, there is no nitrate in the solution. Next, a calcium cation is added to the solution. Only a sulfate anion present in the solution will produce a precipitate with calcium ions. If there is no precipitate, then no sulfate ion is present. The solution left will precipitate only the chloride ion when a silver ion is added. The remaining solution can be tested for the acetate ion by adding an acid. The acetate ion plus a hydrogen ion from the acid will generate acetic acid, which smells like vinegar.

The sequence of reactions involved in a qualitative analysis is sometimes referred to as a *qual scheme*. A chemist would diagram this sequence using vertical lines to represent the addition of a reagent to the solution, and horizontal lines to represent a separation of the solution into solid products and the remaining ions. The solid products and ions remaining in solution are noted in boxes. The scheme for this investigation looks like this:

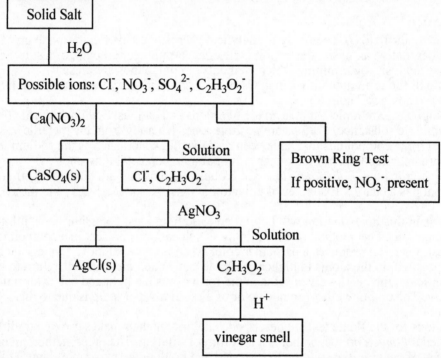

The series of reactions must be followed in order or the results can be confusing. For instance, if $AgNO_3$ is added before $Ca(NO_3)_2$, both AgCl and Ag_2SO_4 can precipitate. The experimenter would not know which precipitated, or if in fact, it was both. For the scheme to work correctly, the SO_4^{2-} needs to be removed before the $AgNO_3$ is added. If a positive test occurs, the rest of the tests do not have to be done.

WASTE AND THE ENVIRONMENT

All the original test solutions can be flushed down the drain with plenty of water, unless instructed otherwise by your instructor. Many scientists are now very cautious about allowing any metal ions into the water systems. The metal ions are not removed by normal sewage treatments, so the metal ions can get into streams, rivers, and the groundwater system. The calcium sulfate, silver chloride, and silver nitrate precipitates can be placed in the trash unless otherwise directed by your instructor. Concentrated acids or bases can damage the plumbing if they are not diluted and/or neutralized before disposal.

▲ *CAUTION*s warn about safety hazards.
*EXTRA*s give helpful hints, additional information, or interesting facts.

Reagents
        1 M chloride salt solutions of
            Na, K, Li, and Cu
        1 M iron sulfate [$FeSO_4$]
        sulfuric acid concentrated [H_2SO_4]
        1 M calcium nitrate [$Ca(NO_3)_2$]
        1 M silver nitrate [$AgNO_3$]
        1 M hydrochloric acid [HCl]
        hydrochloric acid concentrated [HCl]
        unknown salts
        6 M nitric acid (HNO_3) (for silver reclamation)
        1 M sodium hydroxide (NaOH) (for silver reclamation)
        pH test paper (for silver reclamation)

Common Materials
        notebook paper

Laboratory Equipment
        flame test wires
        filter paper (for silver reclamation)
        laboratory burner
        2 50 mL beakers
        1 400 mL beaker
        10 mL graduated cylinders
        small test tube
        1 pair forceps

PROCEDURE
In order to perform a flame test on an unknown cation, you should be able to recognize the color produced by each of the cations. Here, you will test each of the four cation solutions with the flame test and note the characteristic colors of each.

PAPER METHOD
1. Cut widthwise across a piece of notebook paper to produce six strips of paper. Fold the paper in half lengthwise and then again in half lengthwise. This will produce a strip of paper about 8 in. long that is fairly stiff.

2. Light a Bunsen burner. Adjust it to a blue flame. Pour about 10 mL of a salt solution into your smallest beaker. Using forceps, hold one end of a piece of folded notebook paper, dip the other end of the paper in the beaker. Hold the wet end of the paper in the flame of the Bunsen burner. Watch for the color produced as the liquid evaporates. Remove the paper before it catches fire. Rewet the paper as often as needed for both students to see the flame. Use a different piece of paper for each solution. Record the color of each solution on the report sheet.

ALTERNATIVE PROCEDURE USING A FLAME LOOP
3. Light a Bunsen burner. Adjust it to a blue flame.

4. Dip a Nichrome or platinum wire with a loop at the end into 10 mL of concentrated hydrochloric acid and then into the flame. Repeat this process until no color other than that of the flame is apparent. Then dip the wire into 10 mL of the solution to be tested. The color will appear quickly. Repeat the process for each test solution. Record the color of each solution on the report sheet.

Unknown sample
5. Obtain a sample of an unknown salt from the lab instructor. Mark the number of this unknown on the report sheet.

6. Dissolve the salt in 30 mL of water.

Test for Anions
7. Note: When the anion has been identified, record it in the appropriate spot on the report sheet and then skip to procedure 10. Place five drops of the solution in a small test tube for a nitrate test called the brown ring test. Add five drops of iron (II) sulfate ($FeSO_4$) to the test tube. Mix well by swirling. Hold the test tube at a 45° angle. Carefully drop five drops of concentrated sulfuric acid

EXTRA
Fill a 400 mL beaker with water to use to extinguish the paper in case it ignites.

⚠ *CAUTION*
Don't burn yourself.

⚠ *CAUTION*
Don't place burning paper in the trash.

⚠ *CAUTION*
Hydrochloric acid is caustic.
EXTRA
Touching the loop with your fingers will cause a yellow flame because of the salt on your hands.

⚠ *CAUTION*
Sulfuric acid is caustic.

(H_2SO_4) down the side of the test tube. The liquids should form two layers, with the sulfuric acid on the bottom. Set the test tube aside in an upright position for a few minutes. Formation of a brown ring at the interface between the two layers is a positive test for the nitrate ion.

8. Place 10 mL of the unknown salt solution into a 50 mL beaker. Add 5 mL of 1-M calcium nitrate ($Ca(NO_3)_2$). If a precipitate forms, the anion is sulfate.

9. If there is no precipitate, add 5 mL of 1-M silver nitrate ($AgNO_3$) to the solution. If there is a precipitate, the anion is chloride.

10. If there is no precipitate, add 5 mL of 1-M HCl. (The silver added in step 8 will precipitate as silver chloride.) Waft, or fan, the fumes from the test tube to your nose. A vinegar smell is a positive test for acetate ion.

11. **Test for Cations**
    Following the same procedure as earlier, use the remaining salt solution for the flame tests to determine the cation. Use a fresh piece of notebook paper or cleaned flame test wire. When the cation has been identified, record it on the report sheet.

12. An alternative method for disposing of silver is to reclaim it as metal. Dissolve the silver salt in 6-M nitric acid (HNO_3). Neutralize with 1 M sodium hydroxide (NaOH) to pH 7. Place a clean copper strip in the solution. Let it sit for 20 minutes. The silver metal will easily wipe off the copper and into a filter cone (see Investigation 9). Filter the silver metal. Allow it to dry.

EXTRA
Silver nitrate on skin causes a brown stain that will go away in about three days. Avoid skin contact.

⚠ *CAUTION*
Hydrochloric acid is caustic.

⚠ *CAUTION*
Don't burn yourself.

EXTRA
Recycling is better than discarding.

⚠ *CAUTION*
Nitric acid is corrosive.

⚠ *CAUTION*
NaOH is caustic.

FLAME TESTS AND ANALYSIS
PRE-LAB QUESTIONS

Name: _____

Lab Partner: _____

Section: _____   Date: _____

1) Would a flame test work to identify the specific substances in a solution if it contained multiple cations? Explain your answer.

2) What is the advantage of a qual scheme that not only identifies, but simultaneously precipitates, a particular ion?

3) Why would it be important for a technician at a water treatment plant to check the incoming water often for the presence of barium, lead, mercury, or other toxic metallic cations?

4) How might the plant operators deal with the presence of such ions?

5) The chloride ion concentration in ocean water is quite high; nearly 55% of the ions present are chloride. What would you expect to be the fate of any ions of silver that found their way into the ocean? Explain your reasoning.

FLAME TESTS AND ANALYSIS
REPORT SHEET

Name: _____

Lab Partner: _____

Section: _____   Date: _____

I. FLAME TEST COLORS FOR CATION

Sodium   _____

Potassium   _____

Lithium   _____

Copper   _____

Unknown   _____

Unknown   _____

*love procedure
Nature of light*

II. ANION   Positive   Negative

Nitrate test   _____   _____

Sulfate test   _____   _____

Chloride test   _____   _____

Acetate test   _____   _____

III. SALT IDENTITY
    Cation   Anion

_____   _____

IV. QUESTIONS

1. Describe some situations in which an analysis of material is necessary.

H₂O Analyis

2. Is a scheme necessary in those situations?

3. Why is the $Ca(NO_3)_2$ added before $AgNO_3$?

4. What would happen if $AgNO_3$ were added before $Ca(NO_3)_2$? (Hint: Ag_2SO_4 is an insoluble compound.)

5. Why is the brown ring test done on the original solution instead of on the solution left after the sulfate and chloride ions have been removed?

7 Ionic vs. Covalent

Electrons Transferred or Shared

OBJECTIVES

To gain an understanding of the fundamental properties of ionic and covalent compounds.

To classify a group of compounds as ionic or covalent based upon data.

To observe various trends for solubility, electrical conduction, and melting point for ionic and covalent compounds.

Relates to Chapter 4 of Chemistry for Changing Times, *twelfth ed.*

BACKGROUND

There are two basic types of chemical bonds that hold atoms together to form compounds. *Ionic bonds* are formed when one or more electrons are transferred from one atom to another. The resulting positive and negative ions then attract each other, much like the opposite poles of two magnets would attract each other. These bonds are very strong and usually release a great deal of energy in their formation. The strength of ionic bonds can vary with the size and relative charge of the ions involved. Ionic bonds form between any two ions of opposite charge. One ion can form bonds to several nearby oppositely charged ions. In ionic solids, the ions are arranged in a regular pattern, much like a three-dimensional checkerboard. This pattern is referred to as a *lattice* and the energy holding the compound together is *lattice energy*. Ionic substances are generally hard, brittle, crystalline in nature, and possess high melting points. They are also generally water-soluble, and their aqueous solutions (abbreviated "aq") conduct an electric current. We call these solutions *electrolytic*, and the compounds are referred to as *electrolytes*.

Covalent bonds are formed when one or more electrons are shared between atoms. These bonds vary greatly in strength depending, for the most part, on the size of the atoms and the relative attractions of the bonding atoms for the shared electrons. Covalent compounds are generally more soluble in non-polar solvents such as ethanol than in water, have low melting points, and do not conduct electricity well in water solutions. These properties are not as clearly defined in covalent compounds as are the properties of ionic compounds. Water solubility can vary with the polarity of the molecules.

A sample of an ionic compound is many ions held together by ionic bonds. A sample of a covalent compound is many molecules held together by *intermolecular forces*. The molecule itself is held together by covalent bonds, but the forces between molecules are different. They are not true bonds, but are the result of the attraction of negative and positive charges between adjacent molecules. The intermolecular forces are weaker than covalent or ionic bonds. When a covalent compound dissolves or melts, it is the intermolecular forces that are being broken. Because the intermolecular forces are weaker than ionic bonds, covalent compounds have lower melting points and are usually more soluble than ionic compounds. Since a solution of a covalent compound contains uncharged molecules of the compound, it does not conduct electricity well. A few covalent compounds, such as organic acids, are weak electrolytes when in solution.

There are some compounds that appear in a form similar to crystals but are not ionically bonded. The particles in these compounds have a regular geometric arrangement as do ionic compounds, but they are covalently bonded. Diamond is the most famous of these, but quartz and many other gemstones also belong to the class of *covalent networks*. A good portion of the gemstones are silicon- and oxygen-based compounds. These substances are very hard and have high melting points like ionic compounds, but they do not dissolve in either water or non-polar solvents such as ethanol.

By observing the properties of melting point, solubility, and electrolytic capacity for a compound, one can determine the classification of the substance as being either ionic or covalent in nature.

WASTE AND THE ENVIRONMENT

All substances in this procedure except thioacetamide are nontoxic and can be washed down the drain with plenty of water. Thioacetamide should be mixed with a sodium hypochlorite solution and allowed to cool before being flushed down the drain with plenty of water, unless otherwise directed by your instructor.

▲ *CAUTION*s warn about safety hazards.
*EXTRA*s give helpful hints, additional information, or interesting facts.

Reagents
ethanol [CH_3CH_2OH]
distilled water
6 solid unknown compound samples

Common Materials
aluminum foil (or a tin can lid)
pencil

Laboratory Equipment
ring stand
ring
wire gauze
6 small test tubes
microscale conductivity tester
10 mL graduated cylinder
evaporating dish

PROCEDURE
Part A: Melting Point

1. Place a piece of aluminum foil on a flat wire gauze or tin can lid. Turn up the edges of the foil to form a lip. Number 1–6 in a circle that is centered and as close as possible to the outer rim, spacing the digits equally.

2. Obtain the six solid substances from your lab instructor. Number their containers 1–6 to correspond with the six numbers on the foil. Place a pencil eraser-size mound of each substance equally spaced around the circle near the edge of the foil at the place of each number, but not covering the numbers.

3. Describe each substance noting texture, color, translucence, particle size, and shape.

4. Place a lit burner 4 in. beneath the wire gauze. Center the flame so that the circle of substances is centered over it and each pile will be heated equally. Note the time at which each of the piles begin to melt. As soon as the first three have melted, remove the flame.

5. Discard the foil and substances in the trash unless your instructor advises otherwise.

Part B: Solubility and Conductivity

6. Number six small test tubes 1–6 with a pencil on the white spot, or with tape on a test tube rack. Fill each of six test tubes with 3 mL of distilled water.

7. Place a very small amount of one of the substances into its appropriately numbered test tube. Watch to see if it dissolves on the way down. If it reaches the bottom, stopper the tube and shake. If it does not dissolve when shaken, assume it is not water-soluble. Record your results for the water solubility of the substance in the data table.

8. Repeat step 7 for the remaining five substances.

9. Obtain a microscale conductivity tester from your instructor. Pour the contents of one test tube into a clean and dry evaporating dish. Test the solution for electrical conductivity. Record your observations in the data table. Wash and dry the evaporating dish, and rinse and dry the electrodes on the conductivity tester.

10. Repeat step 9 for each of the remaining five test tubes.

11. Rinse the test tubes and dry them. Place 3 mL of ethanol into each test tube.

12. Repeat step 7 for all six substances.

13. Rinse and dry the test tubes.

EXTRA
Translucent means partially transparent.

EXTRA
Three should melt fairly quickly, while the remaining three continue to be in the solid state after 10–15 seconds.

EXTRA
To reduce the risk of contamination and false results, be sure the evaporating dish and conductivity tester are thoroughly cleaned and dried.

EXTRA
The test tubes must be thoroughly dried to obtain accurate results.

⚠ CAUTION
Ethanol is very flammable!

IONIC VS. COVALENT
PRE-LAB QUESTIONS

Name: _____

Lab Partner: _____

Section: _____ Date: _____

1) Table sugar melts at a much lower temperature than does table salt. Which of these two would you most expect to form an electrolytic solution?

2) The labels have fallen off of two containers on a lab counter. The label from one container reads "potassium iodide," an ionic compound. The other container label reads "dextrose," a covalent compound and simple sugar. Describe two tests you could perform to determine how to accurately re-label each container.

3) Ocean water contains many different ions, but the majority of them are of only seven varieties. Sodium and chloride ions together make up approximately 82% of the ions present. What property of ionic compounds is responsible for the presence of these ions?

4) Why would we not expect to find a large amount of silicon dioxide (sand) dissolved in ocean water?

IONIC VS. COVALENT
REPORT SHEET

Name: _____

Lab Partner: _____

Section: _____   Date: _____

I. PART A: MELTING POINT

Substance	Description of Substance	Melting Time
1	_____	_____ s
2	_____	_____ s
3	_____	_____ s
4	_____	_____ s
5	_____	_____ s
6	_____	_____ s

II. PART B: SOLUBILITY AND CONDUCTIVITY

Substance	Water-soluble	Ethanol-soluble	Conductivity of (aq) Solution
1	_____	_____	_____
2	_____	_____	_____
3	_____	_____	_____
4	_____	_____	_____
5	_____	_____	_____
6	_____	_____	_____

III. PART C: CLASSIFYING THE COMPOUNDS

Substance	Ionic or Covalent	Compound Name
1	_____	_____
2	_____	_____
3	_____	_____
4	_____	_____
5	_____	_____
6	_____	_____

57

IV. QUESTIONS

1. What syllable in the name of some compounds is a very good indicator of the class to which it belongs?

2. According to the data gathered in this lab, what is the most definitive property for use in classifying ionic and covalent compounds?

3. Consider the solubility properties of ionic and covalent compounds, and apply this to hair care. When wet hair is rolled and then dried, the curl leaves the hair as soon as it encounters moist conditions again. However, if the hair is treated with a permanent wave, the curl does not wash out. Which type of bond is involved in a "wet set," and which type is involved in a permanent set?

4. Is the statement "Pure liquid compounds are covalently bonded" correct? If so, give three specific examples to support your answer.

8 *Chemical Reactions*

Knowing When Something Happens

OBJECTIVES
To identify changes that indicate a chemical reaction has occurred.
To understand the method of representing a reaction by a chemical equation.

Relates to Chapters 5 and 15 of Chemistry for Changing Times, *twelfth ed.*

BACKGROUND
Chemical reactions occur when a change is made in the bonding of a compound or in the electronic state of an element. Generally, there are visible signs when a chemical reaction has taken place. Even when these are not present, there are certain indications that a reaction is occurring.

Indications that a chemical reaction is occurring include the production of a solid (a *precipitate*), the disappearance of a solid, or the production of a gas. Energy changes can indicate that a reaction has taken place and may include forms of energy such as electricity, light, sound, or heat. Heat may be produced in a reaction, causing a temperature rise, or heat may be absorbed, causing a lowering of the temperature. The color or the pH of a solution may change as a result of a chemical reaction. Regardless of which of these general indicators is or is not present, one statement is always true: Some property will change when a chemical reaction occurs.

Other processes can exhibit indications of a chemical change but do not actually involve an atomic rearrangement or an electron transfer. The process of dissolving a solid in a solvent can cause heat to be produced or absorbed, and evaporation of a liquid produces a gas, but evaporation and dissolution are not chemical reactions because chemical bonds are not changed. However, in a true chemical reaction, other changes will also occur that will be proof that a new substance with new properties has been produced. In other words, we can use certain changes as evidence that there may have been a chemical change, but the only proof that a reaction has occurred is to test the properties of the products to see that they are different from those of the reactants.

Chemists describe reactions by *equations* that serve as a symbolic representation of a chemical process. The beginning substances, or *reactants*, are written on the left, and the substances resulting from the reaction, or *products*, are written on the right.

In this experiment, chemical reactions will be made to occur, providing you a chance to observe the changes that result. The following explanations will help you understand the processes we will observe.

Copper carbonate will not dissolve in water, but in a solution of sodium hydrogen sulfate, the reaction of the hydrogen ion (from the sodium hydrogen sulfate) with the carbonate in copper carbonate will cause it to dissolve. In the reaction of copper carbonate with hydrogen ion, the copper carbonate and the hydrogen ion are reactants. The products are copper ion and bicarbonate ion (hydrogen carbonate ion). The equation for the reaction is written

$$CuCO_3(s) + H^+(aq) \rightarrow Cu^{2+}(aq) + HCO_3^-(aq).$$

Iron atoms will react with copper ions. Two electrons from an iron atom, $Fe(s)$, will transfer to the copper ion (Cu^{2+}) to produce iron ions (Fe^{2+}) and copper atoms, $Cu(s)$. This is an *oxidation-reduction* reaction because of the transfer of electrons. A positive test for iron is the reaction of iron(II) ions with $K_4[Fe(CN)_6]$ to form the blue complex $[Fe_4(Fe(CN)_6)_3]$.

The copper metal reacts with nitric acid to form copper ions (Cu^{2+}), water, and nitrogen dioxide gas. Again, it is an oxidation-reduction reaction. The equation for the reaction is written

$$Cu(s) + HNO_3(aq) \rightarrow Cu^{2+}(aq) + H_2O(g) + NO_2(g).$$

The equation needs to be *balanced*, that is, the same number of atoms of each element must be represented on both sides of the yield sign. The balanced equation is

$$Cu(s) + 4HNO_3(aq) \rightarrow Cu^{2+}(aq) + 2H_2O(g) + 2NO_3^-(aq) + 2NO_2(g)$$

Some of the nitrate ion is not needed in the reaction, so it appears as a product. Other equations used in this investigation are easier to balance.

The dissolution of ammonium sulfide, $(NH_4)_2S$, is not a chemical reaction because there is no change in electron positions within or between the atoms. The ions that make up the compound, ammonium (NH_4^+) and sulfide (S^{2-}), simply become separated and are surrounded by water molecules.

When ions of a compound that is insoluble in water are placed in solution together, solid crystals composed of the ions form. In the precipitation reaction, the copper ion and the sulfide ion react to form solid copper sulfide (CuS). Bonds are formed between the copper ions and the sulfide ions.

The reaction that produces heat is actually an acid-base neutralization reaction. The hydrochloric acid is neutralized by the base sodium hydroxide. The products are water, sodium ions (Na^+), and chloride ions (Cl^-) in solution. This is an *acid-base* reaction because a hydrogen ion is transferred from the acid to the hydroxide ion (OH^-) of the base to form water.

The dissolution of sodium hydrogen sulfate is not a chemical reaction. However, the subsequent gas production is a result of the reaction of hydrogen sulfate ions with the iron atoms to form iron ions, hydrogen gas, and sulfate ions (SO_4^{2-}).

WASTE AND THE ENVIRONMENT

None of the solutions or compounds from this investigation are toxic, although many scientists are now very cautious about allowing any metal ions into the water systems. The metal ions are not removed by normal sewage treatments, so the metal ions can get into streams, rivers, and the groundwater system. The acidic solutions and the basic solutions could cause damage to the plumbing if not neutralized or diluted by a lot of water.

Cyanide is a well-known poison and should not be placed into the water system; however, the cyanide ion is tied up with the iron ion in ferrocyanide, rendering it harmless.

⚠ *CAUTION*s warn about safety hazards.
*EXTRA*s give helpful hints, additional information, or interesting facts.

Reagents

solid copper carbonate [$CuCO_3$]	6 M hydrochloric acid [HCl]
solid sodium hydrogen sulfate [$NaHSO_4$]	6 M sodium hydroxide [NaOH]
solid potassium ferrocyanide [$K_4Fe(CN)_6$]	pH test paper
(or sodium ferrocyanide)	copper turnings
1 M copper sulfate [$CuSO_4$]	6 M nitric acid
1 M potassium iodide [KI]	solid ammonium sulfide [$(NH_4)_2S$]

Common Materials

steel wool, fine

Laboratory Equipment

filter paper
2 small test tubes
2 large test tubes
50 mL beaker
10 mL graduated cylinder
eye dropper
stirring rod

PROCEDURE

1. Color Change

Place 0.1 g of copper carbonate ($CuCO_3$) into a large test tube. Add 0.3 g of sodium hydrogen sulfate ($NaHSO_4$). Add 20 mL of water. Stopper the test tube. Shake to dissolve the solids.

EXTRA
$CuCO_3$ is insoluble in water but will dissolve in a solution of $NaHSO_4$. The blue color is an indication of the Cu^{2+} copper ion.

2.

Add 0.2 g of finely cut steel wool to the test tube and shake. Allow the test tube to stand until the blue color disappears. Filter out the steel wool. Fold a piece of filter paper in half and in half again as shown below. Tear the corner off of the outer fold of the paper. Separate between the first and second folds to form a cone. Place the cone into a glass funnel. Pour the solution through the cone into a beaker. Note the red color on the steel wool.

EXTRA
Steel contains iron. The red color is a deposit of copper metal.

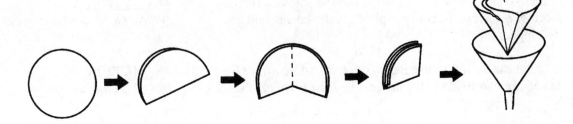

3.

Test the solution for iron (Fe) by adding 0.1 g of potassium ferrocyanide, $K_4[Fe(CN)_6]$, to the beaker. Shake the beaker. The blue color is an indication of iron ions (Fe^{2+}). Pour the solution down the drain with lots of water.

EXTRA
Electrons were forced from the iron metal to the Cu^{2+} ion during the chemical reaction.

Hood
4. Disappearance of Metal

Place 0.5 g of copper turnings into a small test tube. Add 10 mL of 6 M nitric acid (HNO_3). Allow the reaction to continue until the copper metal has disappeared.

⚠ CAUTION
HNO_3 is caustic.
EXTRA
The bubbles are H_2 gas. The blue color of the solution is an indication of Cu^{2+} ions.

No ✗ **Neutralization**

Test the pH of the copper solution by touching the solution with a glass stirring rod and then touching a piece of pH paper. Determine the pH by comparing the color where the drop of solution touched the pH paper with the chart on the pH paper container. Record the pH.

EXTRA
A pH of less than 7 denotes an acidic solution; a pH greater than 7 denotes a basic solution.

6. Add 2 mL of 6 M sodium hydroxide (NaOH) and test the pH. Continue adding NaOH in 2-mL increments until the pH is greater than 7—that is, no longer acidic.

A CAUTION
NaOH is caustic.

7. **Precipitation**
Dissolve 1 g of ammonium sulfide ((NH$_4$)$_2$S), in 10 mL of water in a large test tube. With an eyedropper, add drops of ammonium sulfide to the copper solution. Notice the formation of solid copper (II) sulfide (CuS). Continue to add drops of ammonium sulfide until no more solid forms. The solid will gradually settle, or you can centrifuge it. Notice the color of the solution. The solid can be placed in the trash. Pour the solution down the drain with lots of water.

A CAUTION
It is dangerous to add (NH$_4$)$_2$S to an acid solution.

8. **Heat Production**
Place 10 mL of 6-M hydrochloric acid (HCl) into a small beaker. With a glass stirring rod, touch the acid and then a piece of pH paper. Record the pH. Feel the beaker and note its temperature.

A CAUTION
Hydrochloric acid is caustic.

9. Set the beaker on the lab bench (counter). Add 2 mL of 6-M sodium hydroxide to the beaker. Test the pH. Feel the beaker for warmth.

A CAUTION
NaOH is caustic.

10. Continue adding 6-M NaOH in 2-mL increments until 14 mL has been added. Test the pH and feel the beaker for warmth after each 2-mL addition.

EXTRA
20 drops is about 1 mL.
A CAUTION
Don't burn yourself.

11. Test the pH of the 6-M NaOH. Pour the solutions down the drain with lots of water.

12. **Gas Production**
Place 0.1 g of sodium hydrogen sulfate (NaHSO$_4$) in 20 mL of distilled water in a large test tube. Add a small piece of steel wool. Heat this mixture without boiling for several minutes.

EXTRA
NaHSO$_4$ is also called sodium bisulfate.

13. Test for Fe^{2+} ions in solution by adding 0.1 g of potassium ferrocyanide.

EXTRA
The bubbles are hydrogen gas.

14. Flush the solutions down the drain with plenty of water, and place the solids in a waste container to be buried in a toxic waste dump.

CHEMICAL REACTIONS
PRE-LAB QUESTIONS

Name: _____

Lab Partner: _____

Section: _____   Date: _____

1) A chemical reaction is defined as a process in which the atomic and electronic arrangements of substances are changed. If a substance does not change during a process, should it be represented as a reactant or a product in the equation?

   Catalyst - Neither

2) Many times a substance such as a catalyst is required for reaction, but is not changed in the course of the reaction. Where would be an appropriate place to indicate the presence of the substance in the equation?

3) When testing the pH of a substance, a drop of the substance is added to a piece of pH paper. A color change is the indicator of pH. Could the change in the pH paper itself be a reaction? Support your answer.

4) Some reactions are very subtle and give almost no indicators. Others are so over-the-top that there is no doubt a reaction has occurred. Consider the chemical processes involved in a Fourth of July fireworks display. List all of the reaction indicators that you can think of evident in the explosion of fireworks.

5) Give an example of each of the following changes that would NOT be a chemical change:
   a) temperature change

   b) formation of a gas

   c) formation of a solid

   d) color change

CHEMICAL REACTIONS
REPORT SHEET

Name: _____

Lab Partner: _____

Section: _____   Date: _____

I. COLOR CHANGE
    1. Complete the equation:

       Cu^{2+} + Fe →

    2. Where did the iron ions come from?

II. DISAPPEARANCE OF METAL
    1. What does the color of the solution indicate?

    2. Where did the copper go?

III. NEUTRALIZATION

    Initial pH _____

    Final pH _____

IV. PRECIPITATION
    1. Write the equation for the reaction.

    2. What is the color of the final liquid?

V. HEAT PRODUCTION
1.

	pH
6 M HCl	_____
6 M HCl + 2 mL NaOH	_____
4 mL NaOH	_____
6 mL NaOH	_____
8 mL NaOH	_____
10 mL NaOH	_____
12 mL NaOH	_____
14 mL NaOH	_____
6 M NaOH	_____

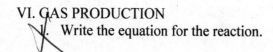

2. Write the equation for the reaction.

3. Why were you instructed to place the beaker on the lab bench (counter)?

VI. GAS PRODUCTION
1. Write the equation for the reaction.

2. Where did the hydrogen come from?

3. Where did the iron ions come from?

9 *Double Replacement Reactions*

Swap Your Partners, Do-Si-Do.

OBJECTIVES

To learn to detect whether a reaction has occurred by observing the four general evidences of a chemical reaction — color change, gas formation, precipitate formation, or energy change.

To use solubility rules to identify solid products from a precipitation reaction.

To use an indicator to identify an acid-base reaction that has no solid or gaseous product.

To correctly complete and balance the net ionic equation representing each reaction that occurred.

Relates to Chapter 4 of Chemistry for Changing Times, *twelfth ed.*

BACKGROUND

Reactions in which two compounds exchange ions to produce two new substances are referred to as *double replacement* reactions. These processes are sometimes called *ionic* reactions or *metathesis* reactions. These reactions usually occur in aqueous environments and are very common reactions in nature and in living organisms. Double replacement reactions follow the general form of AX + BY → AY + BX.

Metathesis reactions are often classified by the products they form. If a reaction produces a solid substance, it is called a *precipitation* reaction. Ionic reactions can also result in gaseous products or the formation of water molecules. Reactions producing water molecules are very often acid-base neutralization reactions that produce a salt at the same time as the water.

Not all combinations of ionic compounds in solution result in a chemical reaction when they are mixed together. Many times all the possible products are water-soluble, and therefore, by the technical definition of a chemical reaction, there is no new substance formed and it can be said that no reaction occurred. When a reaction does occur, it is many times evidenced by a color change, the formation of a precipitate, or the evolution of a gas. However, not all metathesis reactions produce a visible product. Reactions in which water and a soluble salt are formed can be detected by a change in pH using a pH meter or a pH *indicator*. An indicator is a substance that changes color across a specific range of the solution pH.

Many metathesis reactions are isothermic in nature because they do not involve breaking and forming bonds. This is because the ions are already in solution. However, some processes may be endothermic if they produce a gas, or exothermic if an ionic salt is precipitated.

Gas-producing reactions generally result from the addition of an acid to a carbonate or sulfite salt. The resulting weak acid immediately breaks down into water and a gaseous oxide. For example, when carbonic acid (H_2CO_3) is the result of a reaction in solution, it splits to form water and carbon dioxide gas (CO_2). Another gas, sulfur dioxide (SO_2), is the result of the reaction of a metal sulfite and a strong acid.

The results of double replacement reactions are predicted using solubility rules and a basic knowledge of acid-base chemistry. A simple set of solubility rules are given on the next page.

Some double replacement reactions can be written as *net ionic* equations, leaving out the ions that do not actually take part in a chemical change. In net ionic reactions, soluble salts and strong acids are written as ions. Ions that appear on both sides of the equation are then canceled as spectator ions.

Simple Solubility Rules
Note: Apply the rules in the order given. For example, the first rule takes precedent over the remaining rules.

#1 "SNAP" All Sodium, Nitrate, Ammonium, and Potassium salts are soluble.

#2 Silver salts are not soluble (with the exception of silver nitrate, as covered in rule #1, and silver acetate as covered in rule #3).

#3 "ClASH" All Chlorate, Acetate, Sulfate, and Halide salts are soluble (with the exception of silver for chlorate and acetate, [rule #2] mercury, silver, and lead for sulfate and halides [chlorides, bromides, and iodides], and barium and calcium for sulfate).

#4 "CHoPS" All salts containing Carbonate, Hydroxide, Phosphate, and Sulfide are insoluble with the exception of the positive "SNAP" ions (as covered in rule #1) and the additional exceptions of barium, calcium, and magnesium to the sulfide ion.

WASTE AND THE ENVIRONMENT

Solutions from this investigation are not extremely toxic except lead solutions and can be poured down the drain with plenty of water. Most metals are not toxic, although many scientists are now very cautious about allowing any metal ions into the water systems. The metal ions are not removed by normal sewage treatments, so the metal ions can get into streams, rivers, and the groundwater system. The small amount of lead solution used in this experiment can be absorbed by a paper towel and thrown into the trash unless otherwise directed by your instructor. It should not be washed down the drain.

⚠ *CAUTION*s warn about safety hazards.
*EXTRA*s give helpful hints, additional information, or interesting facts.

Reagents
    1 M copper nitrate [$Cu(NO_3)_2$]   1 M potassium iodide [KI]
    1.5 M sodium hydroxide [NaOH]   1 M lead nitrate [$Pb(NO_3)_2$]
    0.5 M hydrochloric acid [HCl]   1 M copper(II) sulfate [$CuSO_4$]
    solid sodium bicarbonate [$NaHCO_3$]   solid sodium sulfite [Na_2SO_3]
    1 M calcium nitrate [$Ca(NO_3)_2$]   1 M sodium chloride [NaCl]
    1 M sodium carbonate [Na_2CO_3]   phenolphthalein
Common Materials
    clear plastic sheet, letter size   white paper
Laboratory Equipment
    10 mL graduated cylinder   droppers for solutions
    100 mL graduated cylinder   50 mL beaker
    3 small test tubes   wash bottle

PROCEDURE
PART A: Precipitation Reactions

1. Using the Solubility Chart provided, predict which combinations of solutions will result in a reaction. Make all predictions before proceeding, and record your predictions in the table for PART A on the Report Sheet. Write "yes" or "N.R." (for no-reaction) in the space provided.

2. Use a piece of white paper inside a clear plastic sleeve as a surface on which to combine the first pair of solutions that is given in the left-hand column of the table for Part A of the Report Sheet. Using only two or three drops of each solution, place the solutions onto the plastic surface in the order given.

3. Observe the result of the combination of solutions. Any clouding or colored powder is a precipitate. Record your observations in the chart on the Report Sheet. Then use the Solubility Chart to identify the precipitate that formed in the spots where a reaction is evident.

4. Repeat steps 2 and 3 for all pairs. Complete one pair of solutions before going onto the next. When finished, wipe the plastic cover dry with a paper towel.

5. Write a balanced equation for each of the reactions on the Report Sheet.

6. Place 2 mL of 1.0 M copper nitrate into a small test tube. Record the temperature. Add 2 mL of 1.5 M sodium hydroxide to the tube. Describe the result, and record the temperature when it has stabilized.

PART B: Gas-Forming Reactions

7. In a 50 mL beaker, obtain 25 mL of 0.5 M hydrochloric acid (HCl). Record the temperature of the acid solution.

8. Place 0.5 g of sodium bicarbonate ($NaHCO_3$) into a small test tube. To that add 3 mL of the 0.5 M hydrochloric acid. Swirl gently to mix. Record your observations on the Report Sheet. When the reaction is complete, record the new temperature of the resulting solution.

9. Repeat step 8 using 0.5 g sodium sulfite (Na_2SO_3) and 3 mL of the 0.5 M HCl. Record your results on the Report Sheet.

EXTRA
To avoid errors due to contamination, be sure not to allow the droppers to touch the reaction surface or any other substances as you combine the solutions.

EXTRA
Do not rinse the plastic cover, or the reaction solutions into sink with the water! Be sure to throw the paper towel away, as lead should not go down the drain. (The paper towel will go into a proper landfill.)

⚠ **CAUTION**
Sodium hydroxide is caustic.

PART C: Acid-Base Reactions

10. Place 2.0 mL of 0.5 M HCl and 1 drop of phenolphthalein indicator into a small test tube.

11. Add 1.0 mL of 1.5 M sodium hydroxide. Record any changes you notice.

DOUBLE REPLACEMENT REACTIONS
PRE-LAB QUESTIONS

Name:

_____

Lab Partner:

_____

Section: _____   Date: _____

1. When solutions of two dissolved salts are combined and result in the formation of an insoluble salt in the form of a precipitate, where are the remaining two ions?

2. Suppose the precipitate in question #1 is filtered from the solution that originally contained four ions. How could the salt that remains in solution be recovered?

3. If there are ions of several soluble salts in one solution, could you devise a way to separate only one of the ions from the solution?

4. Why is the volume of reactants used not important to the results in Part A of this lab?

5. From the procedure in parts B and C, why do you think there are generally cautionary statements on any household products that are moderately or strongly acidic or basic?

DOUBLE REPLACEMENT REACTIONS
REPORT SHEET

Name: _____

Lab Partner: _____

Section: _____   Date: _____

I. PART A: Predictions and Results Based upon Solubility Rules

SOLUTIONS	PREDICTION	OBSERVATION	PRECIPITATE
calcium nitrate & sodium carbonate			
potassium iodide & lead nitrate			
copper sulfate & sodium hydroxide			
sodium chloride & potassium iodide			
sodium carbonate & potassium iodide			
copper sulfate & sodium carbonate			
potassium iodide & sodium hydroxide			
copper sulfate & potassium iodide			

Copper nitrate temperature _____ Copper nitrate and sodium hydroxide temperature_____

II. PART B: Gas-Forming Reactions

	Temperature		Temperature
HCl Solution Observations	_____	NaHCO$_3$ and HCl mixture	_____
Na$_2$SO$_3$ and HCl mixture	_____		

73

Observations

III. PART C: Acid-Base Reactions

Indicator Color

HCl   _____

HCl and NaOH mixture   _____
Observations

IV. QUESTIONS

1. Out of the eight precipitation reactions, how many did you correctly predict? _____

2. In how many of those were you able to correctly determine the precipitate even before observing the reaction? _____

3. Describe the similarities and differences of the precipitates that were formed. Did you find any characteristics consistent with compounds containing a particular ion?

4. Explain why such tools as the Periodic Table, Activity Series, and solubility charts are important to the work of a research chemist.

5. Write a balanced equation for each reaction that occurred in Part A. (If no precipitate formed, list the reactants and write "no reaction" after the yield sign.)

a. _____

b. _____

c. _____

d. _____

e. _____

f. _____

g. _____

h. _____

10 *Iron(II) and Iron(III) Ions*

Red and Blue Switcharoo

OBJECTIVES

To recognize that certain transition elements can exhibit more than one oxidation state.

To understand the concept of an ion as a charged particle and to distinguish between monatomic, polyatomic, metallic, and non-metal ions.

To observe the color differences in compounds containing ions of different oxidation states.

To monitor the varying hues of an ionized transition metal in solution as the concentration of another ion in the solution is raised.

To learn to detect iron(II) and iron(III) ions in solution.

To design a method for idenitfying the presence of both iron(II) and iron(III) ions in the same solution.

Relates to Chapter 4 of Chemistry for Changing Times, *twelfth ed.*

BACKGROUND

Elements combine to form compounds and the "forces" that hold the atoms together are referred to as *chemical bonds*. There are several types of chemical bonds and they can vary greatly in strength. The type and strength of the bonds are major contributing factors to the physical and chemical properties of a compound. This investigation will focus on the peculiarities of a class of compounds containing *ionic bonds* and involving elements known as *transition metals*.

Metallic elements are elements that tend to lose electrons in chemical processes. When one or more electrons are removed from an atom, the atom is no longer neutral because the at-one-time equal number of positive protons and negative electrons is now unbalanced in the favor of the protons. Particles of matter that are not neutral — that is, particles that carry an unbalanced charge — are called *ions*. Nearly all metallic ions are composed of a single atom and are therefore called *monatomic ions*. Many negative ions are also monatomic, such as the oxide ion, O^{2-}.

Ions can be made of more than one atom. When a group of *covalently bonded* atoms carries a single charge, we call the group a *polyatomic ion*. Examples of polyatomic ions are the sulfate ion, SO_4^{2-}, the nitrate ion, NO_3^-, the hydroxide ion, OH^-, and the thiocyanate ion, SCN^-. Polyatomic ions remain intact in many chemical reactions and can therefore be thought of as a single unit. Most polyatomic ions carry a negative charge.

When ions of opposite charge come into close proximity to each other, they are strongly attracted to one another and the ions clump together into a tight arrangement with a regular geometric pattern. These structures are sometimes called *lattices*. These *ionic compounds* (so named because they are composed of charged ions rather than neutral atoms) are hard, brittle, and crystalline in form, and when they contain certain transition metals, often come in bold colors.

Ionic compounds often form in solution when oppositely charged ions can easily come into contact with one another. Ionic substances formed in this manner are called precipitates and are most often seen in a powdered form, causing the solution to temporarily appear to be a cloudy or opaque liquid. With the help of gravity, these small, fine crystals will often settle out of the solution if it is allowed to sit still for a period of time.

Many transition metals have the ability to lose electrons in varying numbers. For example, cobalt will form two ions: cobalt(II), Co^{2+}, and cobalt(III), Co^{3+}. Compounds containing Co^{2+} ions often have a blue hue and the deep blue color of oxide and hydroxide compounds of cobalt, particularly cobalt(II) oxide, is the color commonly called "cobalt blue." Compounds containing the Co^{3+} ion are many times pink in color. There are *complex ions* of cobalt(II) that are blue, such as $[CoCl_4]^{2-}$, and some that are pink, $[Co(H_2O)_6]^{2+}$. A complex ion is a group of ions and/or atoms that include a transition metal ion combined with another ion or small molecule and the group carries a single charge. The grouping is written inside square brackets with the charge indicated on the outside of the bracket. When both of these cobalt(II) complexes are present in solution, a change in temperature, chloride concentration, or water concentration can all shift the dominance to one *specie* or the other and alter the resulting color of the solution.

An iron atom can also lose either two or three electrons when it is *ionized*. The representations for these two ionization processes are:

$$Fe \rightarrow Fe^{2+} + 2\ e^-$$
$$Fe \rightarrow Fe^{3+} + 3\ e^-$$

The iron(II) ion has lost two electrons and subsequently carries a 2^+ charge and therefore has an *oxidation state* of 2^+. The iron(III) ion has lost three electrons and has an oxidation state of 3^+. Iron(II) is found in the complex ion hexacyanoferrate(II), also called *ferro*cyanide. The iron(III) ion is found in the complex ion hexacyanoferrate(III), also called *ferri*cyanide. When either of these complex ions combines with an iron ion of the other oxidation state, the resulting compound is blue in color. These reactions can be used to analyze a solution of iron ions to determine which of the two ions is present.

In this investigation we will observe a solution that contains two different complex ions of cobalt(II). We will note the changes in color that occur with a changing concentration of chloride ions. We will also test iron 2^+ and 3^+ ions with the thiocyanate ion, as well as in different combinations with ferrocyanide and ferricyanide complex ions, to observe the resulting colors.

WASTE AND THE ENVIRONMENT

⚠ *CAUTION*s warn about safety hazards.
*EXTRA*s give helpful hints, additional information, or interesting facts.

Reagents

0.4 M $Co(NO_3)_2$	12 M HCl
0.1 M $FeCl_3$	0.1 M $K_3Fe(CN)_6$
0.2 M $Fe(NH_4)_2(SO_4)_2$	0.1 M $K_4Fe(CN)_6$
0.1 M KSCN	

Laboratory Equipment

1 large test tube	stirring rod
6 small test tubes	test tube rack
watch glass	10 mL graduated cylinder

PROCEDURE
Part A: Cobalt(II) Ions

1. Place 3 mL of cobalt(II) chloride solution ($CoCl_2$) into a large test tube.

2. Record the color in the report table.

3. Add 1 mL of 12 M HCL to the test tube and shake gently to mix the two solutions thoroughly. Record the resulting color.

4. Repeat step 3 until no more color changes are observed.

Part B: Iron(II) and Iron(III) Thiocyanide Compounds

5. Place 1 mL of 0.1 M $FeCl_3$ into a small test tube. Note the color.

6. Add 1 mL of 0.2 M KSCN to the tube and shake gently to mix the two solutions. Record the color of the resulting solution. Set the tube aside to allow the precipitate to settle.

7. Repeat steps 5 and 6 using 0.2 M. $Fe(NH_3)_2(SO_4)_2$ in place of the $FeCl_3$.

Part C: Tests for Iron(II) and Iron(III) Ions

8. Place 1 mL of 0.1 M $FeCl_3$ into a small test tube. Add 1 mL of 0.1 M $K_3Fe(CN)_6$. Shake gently and then note the resulting color on the Report Sheet. Set the test tube aside to settle.

9. Place 1 mL of 0.1 M $FeCl_3$ into a small test tube. Add 1 mL of 0.1 M $K_4Fe(CN)_6$. Shake gently and then note the resulting color on the Report Sheet. Set the test tube aside to settle.

10. Repeat steps 8 and 9 using 0.2 M. $Fe(NH_3)_2(SO_4)_2$ in place of the $FeCl_3$.

11. After the solids in the test tubes have had a chance to settle, pour the clear surface liquid from each test tube into a designated waste container. Use a stirring rod to place a

EXTRA
The pink color is due to the presence of the $[Co(H_2O)_6]^{2+}$ ion.

⚠ CAUTION
12 M HCl is caustic.

EXTRA
As the concentration of Cl^- increases, the Co^{2+} ion converts from the pink $[Co(H_2O)_6]^{2+}$ complex to the blue $[CoCl_4]^{2-}$ complex.

EXTRA
The $FeCl_3$ solution contains Fe^{3+} ions.

EXTRA
The $Fe(NH_3)_2(SO_4)_2$ solution contains Fe^{2+} ions.

EXTRA
$K_3Fe(CN)_6$ contains the Fe^{3+} ion.

EXTRA
$K_4Fe(CN)_6$ contains the Fe^{2+} ion.

77

small amount of each colored substance onto a watch glass. Describe the texture of each substance and the relative sizes of the grains. Note any changes you see as the substances mix.

12. Use a paper towel to collect all insoluble solids and place them into a waste basket or other designated container. Your instructor will dispose of the supernatant liquids.

EXTRA
"Supernatant" refers to the liquid above a precipitated substance that has settled out of solution.

IRON(II) AND IRON(III) IONS
PRE-LAB QUESTIONS

Name: _____

Lab Partner: _____

Section: _____   Date: _____

1) According to the information in this investigation, could a chemist ever be certain of the oxidation state of a cobalt ion from its color alone? Explain.

2) In your own terms, briefly explain the differences in monatomic ions, polyatomic ions, and complex ions.

3) How might it be an advantage to a chemist if the color of a substance was a definite indication of its oxidation state?

4) Which of the following reagents contain polyatomic ions? List the specific ions.
   $Co(NO_3)_2$, HCl, $FeCl_3$, $K_3Fe(CN)_6$, $Fe(NH_4)_2(SO_4)_2$, $K_4Fe(CN)_6$, $KSCN$.

IRON(II) AND IRON(III) IONS
REPORT SHEET

Name: _____

Lab Partner: _____

Section: _____   Date: _____

I. EXPERIMENTAL DATA
PART A: Cobalt(II) Ions

mL 12 M HCl	Color
1	_____
2	_____
3	_____
4	_____
5	_____
6	_____
7	_____
8	_____
9	_____
10	_____
11	_____
12	_____

PART B: Iron(II) and Iron(III) Thiocyanide Compounds

Compound	Color
Fe^{3+} + SCN^-	_____
Fe^{2+} + SCN^-	_____

81

PART C: Tests for Iron(II) and Iron (III) Ions

Iron Ion	ferricyanide (Fe^{3+}) $K_3Fe(CN)_6$	ferrocyanide (Fe^{2+}) $K_4Fe(CN)_6$
Fe^{3+} $FeCl_3$		
Fe^{2+} $Fe(NH_3)_2(SO_4)_2$		

II. QUESTIONS

1. At what point in the addition of HCl to the cobalt solution did the concentration of $[Co(H_2O)_6]^{2+}$ roughly equal the concentration of $[CoCl_4]^{2-}$? How could you tell?

2. What visual result would you expect to see from an increase in the relative amount of water in a solution made of ethanol (alcohol) and water, plus the $[CoCl_4]^{2-}$ and $[Co(H_2O)_6]^{2+}$ ion combination?

3. Why is the thiocyanate ion test only useful for a solution that contains a single iron ion?

4. What test(s) could you perform to prove beyond doubt that a solution contained both iron(II) and iron(III) ions?

5. If a chemist is presented with a solution that is known to contain ionized iron atoms, how many steps are necessary to determine if it contains only one, or both, iron ions? What are they and how would you proceed?

INVESTIGATION

11 *Mole Relationships*

Conserve That Mass

OBJECTIVES

To observe the conservation of mass in a chemical reaction.
To validate mole relationships in a reaction.
To identify the reactant that limits the amount of product formed.

Relates to Chapters 2 and 5 of Chemistry for Changing Times, *twelfth ed.*

BACKGROUND

Chemical reactions occur on a molecular level. The reactions that we observe are actually millions of billions of molecules reacting together. Scientists use the mole unit for calculating amounts of material. One mole is 6.02×10^{23} things, that is, 1 mol of carbon is 6.02×10^{23}, or six hundred two septillion, atoms of carbon.

The molecular weight, or molar mass, of a compound is the sum of the atomic weights of the elements in the compound. The molar mass of CO_2 is calculated by adding the atomic mass of one carbon atom, 12, plus the atomic masses of two atoms of oxygen, each with a mass of 16, for the molar mass of 44. When the unit of grams is used in place of atomic mass units, the *gram atomic mass* of carbon is 12 g and the *gram molar mass* of CO_2 is 44 g. Thus, 1 mol of carbon dioxide molecules (CO_2) has a mass of 44 g.

Chemical equations are written so that the coefficients in the formula represent the relative numbers of moles of reactants and products in the reaction. For example,

$$2H_2 + O_2 \rightarrow 2H_2O$$

Two moles of hydrogen react with 1 mol of oxygen. Expressing molar masses in grams means that 2 moles × 2 g/mole (4 g) of hydrogen reacts with 1 mole x 32 g/mole (32 g) of oxygen to produce 2 moles × 18 g/mole (36 g) of water.

All reactions follow the *law of conservation of mass*. A properly balanced equation correctly predicts the ratio of reactants and products that obey this law. In any reaction, the total mass of the reactants is equal to the total mass of the products. Even if the product of the reaction is a precipitate or a gas, the conservation of mass still holds. If more than 4 g of hydrogen is mixed with 32 g of oxygen, only 4 g of hydrogen will react. The rest of the hydrogen will be excess, and the same amount of water will be produced. The oxygen is the *limiting reagent*. That is to say that it *limits* the reaction because when one reactant—in this case, the oxygen—runs out, the reaction stops and no more product can be produced.

The concentration of a solution is often measured in *Molarity, (mol/L)*. A 1-molar (M) unit is 1 mole of solute per liter of solution. For example, using 1 L of a 2-M solution of is the same as using 2 mol of solute.

We will examine three reactions in this investigation. We will perform a precipitation reaction, a gas-forming reaction, and a reaction to illustrate the effects of a limiting reactant.

In the first reaction, magnesium ions from magnesium chloride ($MgCl_2$) combined with hydroxide ions from sodium hydroxide (NaOH) will produce a precipitate of magnesium hydroxide, $Mg(OH)_2$. A solution of sodium ions (Na^+) and chloride ions (Cl^-) will also be produced as those ions remain un-reacted and dissolved in the solvent.

In the second reaction, the hydrogen ion from acetic acid (CH_3COOH) will react with the bicarbonate ion (hydrogen carbonate, HCO_3^-) from sodium bicarbonate ($NaHCO_3$) to form carbonic acid (H_2CO_3), which then decomposes to form water (H_2O) and gaseous carbon dioxide (CO_2). A solution of sodium ions (Na^+) and acetate ions (CH_3COO^-) is also produced. Since 0.01 mol of reactants is used, 0.01 mole of products will be produced. At room temperature and normal barometric pressure, 0.01 mol of CO_2 is about 230 mL of gas. The balloon used in the procedure needs to be about the same size as a 250 mL beaker.

Calcium sulfate is an insoluble compound in water. In the final reaction, calcium ions and sulfate ions are present together in solution, so they will combine to form a solid (a precipitate). The composition of this solid is one calcium ion to one sulfate ion. If there is an excess of one ion, as in steps 5 and 6, the excess will remain in solution. This will be shown by altering the initial amount of the ion that is used up (the limiting reagent) and then comparing the amount of precipitate formed in each trial.

WASTE AND THE ENVIRONMENT

Although the solutions and solids formed in this investigation are not toxic, it is better not to introduce any metal ions into the water system. It is safer to tie up metal ions in insoluble compounds and bury them in a secure landfill. Concentrated acids or bases can cause damage to the plumbing if they are not diluted and/or neutralized before disposal. Follow the instructions of your instructor for disposal.

⚠ *CAUTION*s warn about safety hazards.
*EXTRA*s give helpful hints, additional information, or interesting facts.

Reagents
    1 M calcium chloride [$CaCl_2$]
    1 M sodium sulfate [Na_2SO_4]
    1 M calcium nitrate [$Ca(NO_3)_2$]
    1 M copper(II) sulfate [$CuSO_4$]
    1 M magnesium chloride [$MgCl_2$]
    1 M sodium hydroxide [NaOH]
    solid sodium hydrogen carbonate, also called sodium
    bicarbonate [$NaHCO_3$]
    1 M acetic acid [CH_3COOH]

Common Materials
    large party balloons

Laboratory Equipment
    filter paper
    Buchner funnel
    centigram balance
    10 mL graduated cylinder
    500 mL Erlenmeyer flask with stopper to fit
    250 mL Erlenmeyer flask
    1 large test tube

PROCEDURE
Part I: Precipitation
1. Place 10 mL of 1 M magnesium chloride ($MgCl_2$) in a 500 mL Erlenmeyer flask. Place 10 mL of 1-M sodium hydroxide (NaOH) in a large test tube that will stand up in the flask. Place a rubber stopper in the flask and weigh it. Record the mass on the report page.

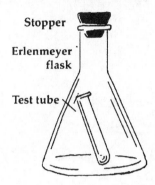

Stopper

Erlenmeyer flask

Test tube

2. Gently invert the flask so the two solutions mix well. Reweigh the flask and record the new mass on the report page.

Part II: Gas Production
3. Place 0.84 g of sodium bicarbonate ($NaHCO_3$) in a 250 mL Erlenmeyer flask. Place 10 mL of 1-M acetic acid (CH_3COOH) in a test tube that will stand up in the flask. Place a balloon over the neck of the flask and weigh it. The balloon should be collapsed when put on the flask. Record the mass on the report page.

4. Gently invert the flask so the two solutions mix well. Do not let the acid touch the balloon. Reweigh the flask and record the mass.

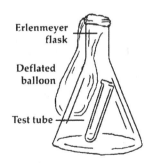

Erlenmeyer flask

Deflated balloon

Test tube

▲ _CAUTION_
Sodium hydroxide is caustic.

EXTRA
The precipitate is $Mg(OH)_2$.

EXTRA
$NaHCO_3$ is also known as baking soda.
▲ _CAUTION_
Acetic acid is caustic.

EXTRA
The gas is CO_2.

Part III: Limiting Reactant

5. For run 1, use a graduated cylinder to measure 10 mL of 1 M sodium sulfate (Na_2SO_4). Place the sodium sulfate in a beaker. Using a clean graduated cylinder, add 10 mL of 1 M calcium chloride ($CaCl_2$) to the beaker. Stir to mix well. Filter by suction through filter paper in a Buchner funnel. After drawing air through the calcium sulfate ($CaSO_4$) precipitate for 5 minutes, weigh the precipitate and filter paper on a watch glass. Record this mass on the report page.

EXTRA
10 mL of a 1 M solution contains 10 mmol of the compound.

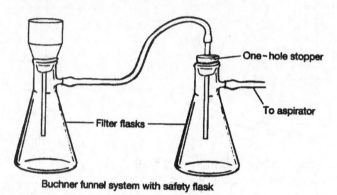

Buchner funnel system with safety flask

6. Divide the filtered liquid from the flask into two beakers. Add 1 mL of 1 M calcium nitrate ($Ca(NO_3)_2$) to one beaker. Add 1 mL of 1 M copper(II) sulfate ($CuSO_4$) to the second beaker. Notice the small amount of precipitation.

7. For run 2, repeat steps 5 and 6 but use 20 mL of 1 M sodium sulfate. In step 6, take note of the amount of precipitate generated from adding calcium nitrate.

8. For run 3, repeat steps 5 and 6 using 20 mL of 1 M calcium chloride with 10 mL of 1 M sodium sulfate. In step 6, take note of the amount of precipitate generated from the addition of copper sulfate.

9. Pour together all the calcium and sulfate solutions. Filter the calcium sulfate precipitate. Discard all precipitates in a container to be buried in a landfill unless directed to do otherwise by your instructor. Pour the remaining solutions down the drain with lots of water.

MOLE RELATIONSHIPS
PRE-LAB QUESTIONS

Name: _____

Lab Partner: _____

Section: _____   Date: _____

1) Refer to procedural steps 1 and 2. How could the sodium chloride (Na^+ and Cl^- ions) that remains dissolved in the solution be recovered?

2) The dissociation of carbonic acid to form water and carbon dioxide gas in step 4 is a major component of the soft drinks we consume. What happens to the acid content of a carbonated drink when the cap is removed and the carbon dioxide gas is able to escape?

3) What remains in solution at the end of step 5?

4) If it was determined that the water supply for a city had a minute level of lead ion contamination, the water treatment plant would likely attempt to eliminate the lead by precipitating it. Would they want the negative ion they add to the water or the existing lead ion to be the limiting reactant? Explain your reasoning.

MOLE RELATIONSHIPS
REPORT SHEET

Name: _____

Lab Partner: _____

Section: _____   Date: _____

I. PRECIPITATION
Mass of flask: before reaction   _____ g

after reaction   _____ g

II. GAS PRODUCTION
Mass of flask: before reaction   _____ g

after reaction   _____ g

III. LIMITING REACTANT
Record the mass of $CaSO_4$ + filter paper for each run:

Run 1   10 mmol Na_2SO_4 +   10 mmol $CaCl_2$ _____ g

Run 2   20 mmol Na_2SO_4 +   10 mmol $CaCl_2$ _____ g

Run 3   10 mmol Na_2SO_4 +   20 mmol $CaCl_2$ _____ g

Precipitation in filtered solution:
Indicate which solution produces a large amount of precipitate.

				$CuSO_4$	$Ca(NO_3)_2$
Run 1	10 mmol Na_2SO_4	+	10 mmol $CaCl_2$	_____	_____
Run 2	20 mmol Na_2SO_4	+	10 mmol $CaCl_2$	_____	_____
Run 3	10 mmol Na_2SO_4	+	20 mmol $CaCl_2$	_____	_____

IV. QUESTIONS
1. What is the relationship between the masses of reactants and the masses of products in the precipitation flask reaction?

2. What is the relationship between the masses of reactants and the masses of products in the gas-production flask reaction?

3. Write the balanced equations for the two flask reactions.

89

4. The three $CaSO_4$ weights should be the same except for a small water-weight difference. Explain this statement.

5. Which compound was in excess in run 2?

6. Which compound was the limiting reagent in run 2?

7. Which compound was in excess in run 3?

8. Which compound was the limiting reagent in run 3?

12 *Molar Volume of a Gas*

Temperature, Pressure, Volume, …Moles!

OBJECTIVES

To produce hydrogen gas from the reaction of a metal and an acid.
To review the concept of an activity series.
To measure the volume and temperature of the gas collected in a eudiometer.
To find the partial pressure of a gas collected over water.
To examine the molar relationships in a chemical reaction.
To calculate the theoretical yield of gas from the mass of metal consumed.
To determine the actual and percent yield of the gas.
To practice using the basic gas laws.
To experimentally determine the standard molar volume of a gas.
To find a percent error on the experimentally determined standard molar volume of a gas.

Relates to Chapter 5 of Chemistry for Changing Times, *twelfth ed.*

BACKGROUND

Gases are an important part of chemistry, but their measurement is not as easy as taking the mass of a solid or the volume of a liquid. Because gases expand to fill any container, four values are required to completely describe a sample of gas. These values are the absolute temperature (K), pressure (P), volume (V), and number of moles (n) of the gas. The four values are related by a *gas constant* symbolized by "R" in the formula: $PV = nRT$, called the *ideal gas law*. If the pressure is expressed in atmospheres, the temperature in Kelvin, and the volume in liters, the value of the gas constant is 0.08206 L·atm/mol·K. It follows then that any gas, regardless of particle size or particle mass, will have an expected volume at a specified set of conditions. The specified conditions are what we refer to as *standard conditions* and they are 0.00 °C (273.15 K) and 1.00 atm pressure. At these conditions, one mole of any gas occupies 22.4 liters of space. This value is known as the *standard molar volume of gases*.

Gas pressures, volumes, and temperatures are further related by the *combined gas law*. This law incorporates both *Boyle's law* and *Charles's law*. These laws predict the volume changes for a constant amount of gas at varying pressures and temperatures, respectively. The combined gas law accurately predicts the new volume of a specific amount of a gas when temperature and pressure are altered. The relationship of these values is $PV/T = P'V'/T'$ where the prime (') indicates the new, or changed, conditions.

Specific pieces of equipment, or arrangements of standard equipment, are required in order to produce and measure a gas. Gases are easiest to collect by liquid displacement. In this experiment, a long calibrated tube called an *eudiometer* will be used. The tube is calibrated for measuring the amount of liquid displaced, or lost, from the tube and is therefore calibrated "for delivery," like a buret. This means the zero is at the top of the tube and it must be read "upside down" so to speak.

If the liquid to be displaced is water, the gas should be one that does not dissolve in the water. Hydrogen is easily produced from an acid and is not water-soluble. For an acid to react to produce hydrogen, we must replace the hydrogen atoms in the acid molecule with another atom. This is a single replacement reaction and the other element will need to be a metal. In looking at the *activity series*, it is evident that metals above hydrogen in the series will be appropriate for the reaction. This exercise will use magnesium ribbon because it reacts easily and fairly quickly with most acids. The acid of choice will be hydrochloric (HCl). The by-product will be magnesium chloride ($MgCl_2$) and it will remain dissolved in the solution. The gas that is produced will rise and its pressure will force the water out through the hole in the stopper at the bottom of the eudiometer.

The measurements of the collected gas will be taken once it has had a chance to sit a while to be certain all hydrogen has left the solution, the temperatures have stabilized, and the water vapor in the eudiometer has reached its equilibrium pressure. The pressure of the collected hydrogen gas can be found in two steps. The first step requires equalizing the gas pressure inside the tube with the atmospheric pressure in the room. This is done by matching the water levels inside and outside the eudiometer. The eudiometer will be placed into a tall cylinder of water and raising or lowering it until the two levels are at the same point on the eudiometer. The second step is to "correct" the gas pressure by subtracting the pressure of the water vapor from the total pressure. *Dalton's Law of Partial Pressures* states that the total pressure in a mixture of gases is equal to the sum of the pressures of the individual gases. Since we will collect the hydrogen over water, the second gas present in the top of the tube will be water vapor. Published tables list the equilibrium vapor pressure of water for specific temperatures.

The volume will be read directly from the eudiometer at the point where the water levels inside and outside match and the temperature will be the room temperature as measured in the water surrounding the eudiometer tube.

We can get a picture of the efficiency of the reaction and of our abilities to perform it by calculating a *percent yield*. The *theoretical yield* of the hydrogen gas is a prediction of the amount we should be able to collect based on *stoichiometry*, or mole ratios, presented in the balanced equation. If it were a perfect world, we could get this amount. The percent we get of the calculated amount is the percent yield.

We can also calculate a *percent error* on the value we determine experimentally for the standard. molar volume of a gas. The percent error is the absolute difference between the accepted and the experimental values times 100% divided by the accepted value.

In this experiment, we will react magnesium metal with a dilute solution of hydrochloric acid and collect the hydrogen gas over water in a eudiometer. From measurements of pressure, temperature, and volume of the collected gas, we will calculate the number of moles of gas produced.

$$Mg(s) + 2\ HCl(aq) \rightarrow MgCl_2(aq) + H_2(g)$$

WASTE AND THE ENVIRONMENT

The solution from the eudiometer can be washed down the drain with plenty of water. Place the thread in the trash can and rinse the eudiometer tube with water twice and with distilled water once. Return it to the instructor.

⚠ *CAUTION*s warn about safety hazards.
*EXTRA*s give helpful hints, additional information, or interesting facts.

Reagents
    magnesium ribbon   6.0 M hydrochloric acid [HCl]

Common Materials
    sand paper   thread
    ruler (cm)

Laboratory Equipment
    50-mL eudiometer   1-hole rubber stopper to fit eudiometer
    10-mL graduated cylinder   wash bottle with distilled water
    1000-mL graduated cylinder   600-mL beaker
    ring stand   utility (or buret) clamp
    barometer   thermometer
    centigram balance

PROCEDURE

1. Fill a 1-liter graduated cylinder with tap water and place it in the sink to allow it to come to room temperature.

2. Fill a 600-ml beaker half full of tap water and place it on a ring stand. Place a utility clamp or buret clamp on the stand so that it sits about 6 inches above the beaker.

3. Cut a piece of magnesium ribbon that is about 4.5 centimeters long. Use sandpaper to buff the ribbon until it is all shiny and there is no dark magnesium oxide (MgO) visible on the ribbon.

4. Measure the mass of the ribbon to the nearest 0.01 gram. If the mass is over 0.040 grams, discard some of the magnesium in the trash and re-measure the mass.

5. Roll the ribbon into a loose coil and use one end of a 12-inch piece of thread to tie it.

6. Use the graduated cylinder to measure 10 mL of 6.0-M hydrochloric acid (HCl) and add it to the eudiometer.

7. Using the distilled water in the wash bottle, hold the eudiometer at a 45° angle and slowly send a stream of water down the side so that there is little turbulence and mixing of the water with the more dense acid solution. When you have added 10 milliliters or so, hold the eudiometer vertically to add enough distilled water to fill the tube.

8. Place the magnesium coil into the top of the tube and use a stirring rod to push it under the water to a depth of about 5 cm. Refill the eudiometer with distilled water if necessary.

9. Allowing the end of the thread to hang over the lip of the eudiometer, gently place the one-hole stopper into the tube to hold the thread firmly in place. Be certain that water comes through the hole as the stopper is gently wedged in place and that there are no bubbles in the top of the eudiometer.

10. Holding a gloved finger over the hole in the stopper, invert the eudiometer tube into the beaker of water so that the bottom 4 cm of the tube is immersed. Release your finger and clamp the tube into place. Make observations as the denser acid moves down the tube and comes into contact with the magnesium metal. Make note if the bubbles are moving into the water in the beaker from the opening in the stopper at the bottom of the eudiometer tube. This will be important to your error analysis.

11. When the reaction is complete, keeping the eudiometer in the water, tap the eudiometer tube gently to release any bubbles of gas that are clinging to the side of the tube or to the stopper. To release the bubbles on the thread, gently pull the thread up parallel to the tube until only the loop and knot remain in the tube. (It might be a good idea to hold the stopper in place as you do this.)

EXTRA
The magnesium oxide is a contaminant and will not react with the hydrochloric acid.

⚠ CAUTION
The acid is corrosive.

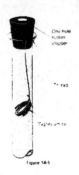

Figure 14-1

EXTRA
If there are bubbles, remove the stopper, refill the tube and try again.

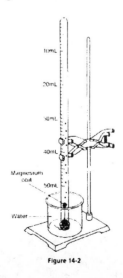

Figure 14-2

12. Holding a gloved finger over the hole in the stopper, gently transfer the eudiometer to the tall cylinder of water in the sink. Lower it until it rests on the bottom of the cylinder. Allow it to sit for five minutes.

13. Raise the eudiometer tube slowly until the liquid level inside the tube is even with the water outside the tube. At this position, read the meniscus of the liquid level inside the tube and record it in the data table.

EXTRA
Did you notice what happened to the water level in the tube as you raised it?

14. Take the temperature of the water in the cylinder and record it in the data table.

15. From a barometer in the room, record the barometric pressure in the data table.

16. Remove the stopper from the eudiometer and discard the magnesium chloride solution down the drain, chasing it with lots of water. Throw away the thread. Rinse the tube once with tap water and once with 10 mL of distilled water, rolling the tube to rinse all surfaces.

17. Repeat the experiment two more times to get measurements for three trials total.

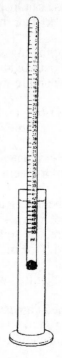

MOLAR VOLUME OF A GAS
PRE-LAB QUESTIONS

Name: _____

Lab Partner: _____

Section: _____   Date: _____

1) If gases expand to fill any container, what invisible barrier keeps an atmosphere around the earth?

2) If the gas collected dissolves in water, as does oxygen or carbon dioxide, what effect will that property have on the final experimental value for the volume of gas produced?

3) A eudiometer is marked in graduations of 0.1 mL. If a water meniscus sits on the third mark above the 46-mL mark, what is the volume reading?

4) If bubbles of the gas are allowed to remain on the inner surface of the tube and on the thread, what effect will that have on the measured volume of gas in the tube?

MOLAR VOLUME OF A GAS
REPORT SHEET

Name: _____

Lab Partner: _____

Section: _____   Date: _____

Data Table

	Trial 1	Trial 2	Trial 3
Mass of magnesium (g)			
Gas volume (mL)			
Temperature ($^{\circ}$C)			
Barometric pressure (in$_{Hg}$)			

Calculations

1) Find the moles magnesium consumed by dividing the mass of magnesium by its molar mass (MM$_{Mg}$ = 24.31 g/mol). Record these values in the calculations table.

2) Convert your volumes in mL to liters by dividing by 1000mL/L. Record these values in your calculations table.

3) Your thermometer is probably calibrated in degrees Celsius. Convert the experimental temperatures to Kelvin by adding 273.15. Record these values.

4) If your barometer is calibrated in inches of mercury, convert the measurement to centimeters of mercury by multiplying by 2.54 cm/in. Then rewrite the measurement in millimeters of mercury, "mmHg," by multiplying by 10. Record this value in the calculations table.

5) Find the experimental temperatures on the table of equilibrium vapor pressures for water and subtract the partial pressure of the water to get the "dry" pressure of the hydrogen gas for each trial. Record these values.

6) Convert the dry gas pressures from torr (mmHg) to atmospheres (atm) by dividing the value by 760 mmHg/atm. Record these values.

7) Use the relationship $n = (PV)/(RT)$ where "P" is the corrected pressure in atm, "V" is the experimental volume in liters, "T" is the Kelvin temperature, and "R" is 0.08206 L·atm/mol·K to find the moles of hydrogen gas collected. Record these values.

8) Examine the balanced equation for the reaction between magnesium and hydrochloric acid in the experimental background material. There is a 1:1 mole ratio of magnesium consumed to hydrogen gas produced. Using your determination of the moles magnesium consumed in the reaction as the theoretical, or expected, yield "T," and the moles hydrogen gas actually produced as "A," determine the percent yield for each trial using $A/T \times 100\% = \%$ Yield

9) Use the Combined Gas Law to find the equivalent volume of hydrogen gas at standard conditions. Use the relationship $V' = (PVT')/(TP')$, where P, V, and T are your experimental values and T' is 273 K, while P' is 1.00 atm. Record this as the volume of a gas at standard conditions (standard volume).

10) Using a ratio, find the experimental volume of one mole of hydrogen gas at standard conditions (Exp. SMV), where EV is your experimental standard volume of hydrogen, and E_{mol} is the experimental moles of hydrogen as follows:
$$EV/E_{mol} = SMV/1.00\,mol \text{ or } (EV/E_{mol})/1.00\,mol = SMV$$

11) Using 22.4 L and the accepted value "A" and your Exp. SMV as the experimental value "E," find a percent error for your value of the standard molar volume of a gas using the relationship: $\%$ Error $= |A-E|/A \times 100\%$ where $|A-E|$ is the absolute value of the difference in the accepted and experimental values.. Record this value in the table.

Calculations Table

	Trial 1	Trial 2	Trial 3
Moles of magnesium (mol)			
Gas volume (L)			
Temperature (K)			
Barometric pressure (mmHg)			
Vapor pressure of water (mmHg)			
Dry pressure of hydrogen (mmHg)			
Dry pressure of hydrogen (atm)			
Moles hydrogen collected (mol)			
% Yield for hydrogen gas			
Standard volume of hydrogen (L)			
Experimental SMV of a gas (L)			
% Error for SMV of a gas			

Post-Lab Questions

1) Why was it necessary to tie the magnesium and pin it against the mouth of the eudiometer with the thread? What does this say about the relative densities of the metal and the water?

2) When the eudiometer was being raised in the large cylinder to match the water levels inside and outside, what happened to the inner water level? Use gas pressures to explain this observation.

3) Examine the table of equilibrium vapor pressures for water. Assume all measurements for the hydrogen gas were kept constant except for the room temperature. Would an increase of 5 degrees have meant the presence of more or less hydrogen gas?

4) Give an explanation for the fact that some hydrogen bubbles could actually escape *down* through the hole in the stopper.

5) List as many sources of error as possible in the data for your experimental value for the moles hydrogen produced and for the determination of the standard molar volume of a gas.

13 *Hydrated Compounds*

Getting Water from a Rock

OBJECTIVES

To relate the idea of hydrated and anhydrous crystals to common experience.

To measure the amount of water contained in a hydrated salt crystal.

To determine when all the water has been driven from a crystal by both visual evidence and by heating to a constant mass.

To make observations that confirm the loss of water during the heating process.

To measure grams of anhydrous substance and grams of water driven from the substance by gravimetric means.

To derive the empirical formulas for hydrated salts from experimental data.

To observe the energy changes that occur as water is added to an anhydrous salt.

To understand the concept of composition stoichiometry.

To calculate percent error for experimental data.

Relates to Chapter 4 and 5 of Chemistry for Changing Times, *twelfth ed.*

BACKGROUND

There are a lot of different types of solutions in chemistry. One of these is a liquid-in-solid solution. Many of the ionic compounds we think of as solid crystals in actuality contain quite a bit of water. These compounds are referred to as *hydrates* and the moisture content is called the *water of crystallization*. Compounds that have no water in their crystals are *anhydrous* solids. Some anhydrous compounds will easily incorporate water molecules from the environment into their crystals and in this way become hydrated. Compounds that exhibit this property are called *hygroscopic* and can be used as *desiccants*. A desiccant will keep the moisture content of nearby air very low. You might find a small package of desiccant in a box of new shoes, or in new luggage or purses that may contain leather or other materials that will be damaged by high humidity. At some time you may have opened a sugar bowl to find the crystals have stuck to each other. The tiny crystal surfaces became bonded together when they had mutual attraction to and then subsequently incorporated water molecules from the air. A few compounds are so efficient at the process of taking moisture from the environment that they will literally dissolve in water molecules drawn from the surrounding air. These are *deliquescent* substances and what started as a solid on the lab bench can very soon become a puddle. An example of this is solid sodium hydroxide.

Most hydrated crystals result when molecules of water are included into a crystal as it forms from an aqueous solution. These solid water/ion solutions are *homogeneous*, that is, the water molecules and ions are evenly distributed within the crystal. The water molecules take up specific positions in the geometric *lattice* of the crystal as it forms and the resulting solid is more stable. The ratio of water molecules to the formula unit of the salt is consistent so that hydrated crystals follow the *law of constant (or definite) composition*. These crystals have a well-defined stoichiometry between the ions and water molecules. In other words, the mole ratio of formula units, or of ions, to water molecules will be constant for all samples. This ratio is an example of *composition stoichiometry*, or the ratio of the components that make up a substance. Examples of some common hydrates are magnesium chloride hexahydrate ($MgCl_2 \bullet 6H_2O$), calcium chloride monohydrate ($CaCl_2 \bullet H_2O$), and sodium sulfate decahydrate ($Na_2SO_4 \bullet 10H_2O$). The expression "$MgCl_2 \bullet 6H_2O$" tells us that there are six moles of water for every mole of magnesium chloride in the crystal. The whole number coefficient before the water symbol is sometimes referred to as a *hydrate number*. We can also think of it as six water molecules for every magnesium ion and two chloride ions.

In most cases, the water can be driven from a hydrated crystal simply by heating it. The heat required to remove the water is called the *heat of hydration*. The heat of hydration is a measure of the amount of energy that holds the water molecules in their lattice positions. As the water is removed, the energy content of the crystal increases. This is because crystals that naturally incorporate water do so because they are more stable, that is, they have a lower energy content, in the hydrated condition. If the water is removed, the anhydrous crystals must be stored in airtight containers so that they do not re-hydrate themselves using moisture from the air.

Some compounds exhibit a change in visual properties between the hydrated and anhydrous forms. One of these is copper sulfate. In its hydrated form it is a beautiful cobalt blue crystal, often called blue stone. Once the water is driven out of the crystal, however, it is a dull gray substance with only a hint of blue to it. Some compounds exhibit very little difference in their appearance between the hydrated and anhydrous forms. An example of this is barium chloride.

In many experimental procedures a value we refer to as *percent error* is a valuable tool in determining how well the laboratory data fits the accepted data. This value is found by dividing the absolute value of the difference in the experimental and accepted values by the accepted value and then multiplying the ratio by 100%. Because the stoichiometry of hydrates is extremely consistent, this experiment is an excellent place to use the concept of percent error.

In this exercise, we will use heat to drive the water from two different hydrates: copper sulfate and magnesium sulfate. We will use a visual determination to know that we have removed all the water from the copper sulfate, but we will rely on successive mass measurements to know that we have removed all the water from the magnesium sulfate.

WASTE AND THE ENVIRONMENT

Adding metal ions to the aqueous environment is never a good idea. The solids from this experiment are not hazardous and can be disposed of in the trash unless otherwise directed by your instructor.

⚠ *CAUTION*s warn about safety hazards.
*EXTRA*s give helpful hints, additional information, or interesting facts.

Reagents
    hydrated copper(II) sulfate crystals
    hydrated magnesium sulfate crystals

Special Equipment
    evaporating dish (2)
    watch glass (2)
    centigram balance
    laboratory burner and tubing
    desiccator
    tongs
    eye dropper
    ring stand with ring and wire gauze

PROCEDURE

Part A: Dehydration of Magnesium Sulfate

1. Assemble a ring stand, ring, wire gauze, and burner so that the ring is about 2 inches above the top of the burner.

2. Clean and dry an evaporating dish. Place the evaporating dish with watch glass on top on the wire gauze and heat it for three minutes to drive off any volatile substances. Allow it to cool to room temperature.

3. Using clean tongs, move the dish and glass to the balance and record the mass to the nearest 0.01 grams. Once the mass of the dish and watch glass have been measured, use only clean tongs to handle the dish and glass.

4. Place about 3.0 g of hydrated magnesium sulfate crystals into the evaporating dish and replace the watch glass. Measure and record the exact mass to the nearest 0.01 grams.

5. Using clean tongs, carefully move the dish with the salt and glass to the wire gauze.

6. Heat the dish slowly at first, moving the burner back and forth under it to warm it evenly. After 10 to 15 seconds, set the burner beneath the dish and allow it to be heated strongly for 10 minutes and then remove the burner. Observe the underside of the watch glass for signs of a reaction. Using the tongs, move the dish and the salt to a desiccator to cool. (While the evaporating dish assembly is cooling, go to steps 11 through 13.)

7. When the assembly has reached room temperature, use the tongs to move the assembly to the balance and measure the mass to the nearest 0.01 gram.

8. Again move the dish/salt/glass assembly to the wire gauze and heat it for 5 more minutes. Use the tongs to move it to the desiccator to cool.

9. When cool, again measure the mass and record it to the nearest 0.01 gram. If the mass has not decreased from the last measurement, it can be assumed that all the water has been removed from the hydrate. If there is a mass difference, repeat steps 7 and 8.

10. Once the mass of the anhydrous salt has been determined, the salt should be deposited into the designated container for disposal unless otherwise directed by your instructor.

Part B: Dehydration of Copper Sulfate

11. Place a second evaporating dish and watch glass on the wire gauze and heat them for three minutes to remove any volatile substances. Allow them to cool until they are room temperature. Use clean tongs to move the dish and watch glass to the balance and record the mass to the nearest 0.01 gram.

EXTRA
If a hot object is placed on a balance pan, not only could the balance be damaged, but the mass will not be accurate. The heat from the object will warm and expand the air above the pan, creating a buoyant force from the denser air below that helps to support the pan.

12. Add about 3.0 grams of hydrated copper(II) sulfate to the dish and replace the glass. Measure the mass and record it to the nearest 0.01 grams.

13. Using the tongs, move the dish/salt/glass system to the wire gauze and heat it strongly. Observe the underside of the watch glass for signs of a reaction. As the water molecules are driven from the salt, color changes will occur, indicating the amount of water remaining in the crystal. When the brilliant cobalt blue has faded to a lighter, aqua blue, some of the water has been driven from the crystal. There is a specific ratio of water molecules to copper(II) sulfate in this aqua-colored substance. With continued heating, the aqua color will leave and the substance will be a dull gray color. When all of the salt has become gray, continue to heat for two more minutes to be certain that all the water has been removed.

14. Use the tongs to move the dish, salt, and glass to the desiccator to cool to room temperature. Then measure and record its mass.

15. Remove the watch glass and set the evaporating dish with the anhydrous copper(II) sulfate into the sink. Using a dropper, add a few drops of water to the substance and observe the result. Record your observations.

16. The copper(II) sulfate should be placed in the designated container.

17. Clean all equipment and return it to its proper place.

HYDRATED COMPOUNDS
PRE-LAB QUESTIONS

Name:_____

Lab Partner:_____

Section: _____   Date:_____

1) If heat is required to remove the water molecules from hydrated crystals, what would you expect to be the result of dropping water onto a dehydrated crystal?

2) Salt allowed to sit undisturbed in a shaker can "cake" or become stuck together. This happens because most salt contains a bit of calcium chloride and magnesium chloride as minor impurities. Explain what is happening.

3) Once the heating of a mass of hydrated crystal is completed, it will be placed into a desiccator to cool. From the name of the device, what do you think is its purpose?

4) In the bottom of the desiccator is a solid substance. Sometimes this substance is a commercial product called Drierite®, but sometimes it is anhydrous calcium chloride. What is the function of the substance in the bottom of the desiccator? Name another substance that could be used.

5) Does the substance in the bottom of the desiccator necessarily need to be discarded once it has fully hydrated? What is an alternative?

6) The manager of a laboratory stockroom might prefer to have the hydrated form of a crystal rather than the anhydrous form. What might be a practical reason for that preference?

HYDRATED COMPOUNDS
REPORT SHEET

Name: _____

Lab Partner: _____

Section: _____   Date: _____

Parts A & B Data Table

	Magnesium Sulfate	Copper(II) Sulfate
Appearance before heating		
Appearance after heating		
Observations during heating		
Mass of dish and glass		
Mass of dish, glass, and salt before heating		
Mass of dish, glass, and salt after first heating		N/A
Mass of dish, glass, and salt after final heating		

Parts A & B Calculations
1) For each salt, calculate the mass of the *hydrated salt* by subtracting the mass of the dish and glass from the total mass before heating. Record these values in the Calculations Table on the next page.

2) For each salt, calculate the mass of the *anhydrous salt* by subtracting the mass of the dish and glass from the final mass after heating. Record these values.

3) For each salt, calculate the mass of water removed from the salt by finding the difference in the masses of the hydrated and anhydrous forms of the salt. Record these values.

4) For each salt, find the moles of anhydrous salt by dividing the mass of anhydrous salt by its molar mass. Record these values. (The molar mass of $CuSO_4$ is 159.60 g/mol, and the molar mass of $MgSO_4$ is 120.38 g/mol.)

5) For each salt, find the moles of water that were removed by dividing the mass of the water removed by its molar mass (18.0152 g/mol.) Record these values.

6) For each salt, find the ratio of moles salt to moles water in the hydrate by dividing the moles water by the moles of anhydrous salt. Record these values expressed to the nearest .01 and in the form "1.00: __."

7) Find the percent error for your experimental values by the following formula:
$$\% \text{ Error} = |A\text{-}E| / A \times 100\%$$
where "A" is the accepted value and "E" is the experimental value for the hydrate number. The accepted hydrate values for magnesium sulfate and copper sulfate are 7.00 and 5.00, respectively. Record the percents error in your calculations table.

8) Round the mole ratio to the nearest whole number to get a hydrate number and report your experimental empirical formula for the hydrated salt in the form of XY·nH_2O where "n" is the experimental hydrate number.

9) Gather hydrate numbers from several classmates as comparison to your work.

Part A & B Calculations Table

	Magnesium Sulfate	Copper Sulfate
mass of hydrate		
mass of anhydrous salt		
mass of water removed		
moles of anhydrous salt		
moles of water removed		
ratio of moles salt to moles water		
% Error		
hydrate number		
exper. hydrate formula		

Post-Lab Questions

1) Once the initial mass of the clean and dry evaporating dish and watch glass have been recorded, they are handled only with clean tongs. Why is this important?

2) What is the purpose of the watch glass?

3) If the final mass is taken while the dish is still hot, what would be the likely effect on the experimental ratio of moles water to moles salt? Would it be too high or too low? Explain the logic of your reasoning.

4) Examine data from some classmates. Does there seem to be consistency in the hydrate numbers you found for the two salts? Comment. What law does this data support?

5) Why was the magnesium sulfate heated until the mass was constant?

6) A student runs an experiment and heats a white magnesium sulfate ($MgSO_4$) crystals and measures the masses before and after heating. During the process, she notices condensation on the watch glass covering the substance. The masses she records are 18.49 and 9.03 grams.
   a) What three evidences are there that the substance was a hydrated crystal?

   b) Calculate the hydrate number for the substance.

7) From your observations when adding water to the copper(II) sulfate, explain the energy changes taking place.

8) Cobalt(II) chloride has an interesting property. A $CoCl_2 \cdot 4H_2O$ is blue, while $CoCl_2 \cdot 6H_2O$ is pink. Can you think of a good use for this substance?

14 *Solubility*

Vinegar and Oil Separate

OBJECTIVES

To understand and predict which liquids will be miscible.
To discover the role of varying viscosity in the mixing of two liquids.
To measure mass and volume and then calculate density for several liquids.
To observe the use of soap as both an emulsifying agent and a cleaning agent.

Relates to Chapter 6 of Chemistry for Changing Times, *twelfth ed.*

BACKGROUND

Molecules similar to each other in shape and polarity will usually dissolve in each other—that is, they are *miscible*. Water is a polar molecule that is usually *immiscible* with nonpolar oils. Vinegar-and-oil salad dressing is an example of two immiscible liquids. The more dense solution of water and acetic acid, which we call vinegar, forms a separate layer beneath the less dense oil layer.

To blend these immiscible components into one liquid called an *emulsion* requires an emulsifying agent. The emulsifier works because one end is soluble in water while the other end is soluble in oil. Thus, the water and oil molecules are held together. Eggs are used as an emulsifying agent in mayonnaise. In an older jar of mayonnaise, a top layer of oil often can be observed because some of the oil has separated.

Two liquids may appear immiscible if one is very viscous or thick. *Viscosity* is the ability to flow. A very viscous liquid will flow very slowly and thus will inhibit mixing and miscibility. Syrup and water will be used as an example of varying viscosities in this investigation. The syrup and water can be forced to mix by putting energy into the system. The energy is added by stirring vigorously.

Liquids that are miscible may dissolve in each other so well that the new volume is less than the sum of the individual volumes. The molecules fit among one another so well that they require less space together than by themselves.

Another aspect of solubility is that one set of intermolecular attractions may be stronger than another. In this investigation the attraction of oil for cloth will be stronger than the attraction of gum to cloth. In this way, the oil will decrease the gum-to-cloth attraction and allow the gum to be removed.

Soap works in much the same way to remove soil from cloth or other materials. The soil-to-cloth attraction is broken by the stronger attraction of soap to soil. Soap acts as an emulsifying agent. Soap molecules have a long organic end that is soluble in oil or grease and an ionic end that is soluble in water. The organic ends of soap molecules are dissolved in small blobs of oil. The ionic ends are then left sticking out to attract water. The entire mass is called a *micelle*. Watch the soap commercials on television that show dirt leaving a dish or cloth. The dirt is rolled into a ball.

WASTE AND THE ENVIRONMENT

The solutions used in this investigation are not toxic.

⚠ *CAUTION*s warn about safety hazards.
*EXTRA*s give helpful hints, additional information, or interesting facts.

Reagents

alcohol (ethyl or isopropyl)

Common Materials

syrup   chewing gum
cooking oil   peanut butter
liquid soap   cloth

Laboratory Equipment

balance   stirring rod
50 mL beaker   10 mL graduated cylinder
eye dropper

PROCEDURE

1. Weigh a clean, dry 10 mL graduated cylinder. Fill the cylinder with exactly 10 mL of syrup and reweigh. Pour the syrup into a 50 mL beaker and set it aside.

2. Wash the cylinder, dry the outside, fill it with exactly 10 mL of water, and reweigh. Pour the water very carefully down the side of the beaker so that the water does not mix with the syrup. Set the beaker aside.

3. Dry the cylinder inside and out. Fill the cylinder with exactly 10 mL of cooking oil and reweigh. Pour the oil carefully down the side of the beaker so the layers do not mix. When you finish, you should have three separate layers.

4. With a stirring rod, mix the three layers well and set the beaker on the lab bench. Two layers will re-form.

5. With an eyedropper, remove enough of only the top layer to put exactly 5 mL into the 10 mL graduated cylinder. Weigh the cylinder and liquid. Record the masses on the report sheet. Clean and dry the cylinder.

6. With an eyedropper, reach through the solution and remove enough of only the bottom layer to put exactly 5 mL into the 10 mL graduated cylinder. Weigh the cylinder and liquid. Record the masses on the report sheet.

7. Add 1 mL of soap to the beaker. Stir well and set the beaker on the lab bench. Report observations on the report sheet.

8. Place exactly 5.0 mL of water in a 10 mL graduated cylinder. In a second 10 mL cylinder, place exactly 5.0 mL of alcohol. Pour the two liquids together in the water cylinder. Stir them together and then read the new volume.

9. Chew a piece of gum to remove most of the sweet taste. Place the gum in a cloth square and squeeze to stick the gum to the cloth. Take a small scoop of peanut butter and rub it into the cloth, loosening the gum from the cloth. When the gum is removed, wash out the peanut butter with soap.

   EXTRA
   Peanut butter will also remove gum from hair.

10. Pour the solutions down the drain with plenty of water. Throw the gum in the trash.

SOLUBILITY
PRE-LAB QUESTIONS

Name: _____

Lab Partner: _____

Section: _____   Date: _____

1) If the liquid layers in the procedure were added to the cylinder in the reverse order—that is, with the least dense first and densest last—what would you expect to happen at the interface of each liquid?

   Would mix = hard to accom

2) What properties would you expect the molecules of two miscible liquids to have in common?

   polar

   non polar

3) Suppose a garment has been stained by a substance that is made of non-polar molecules. What kind of molecule would you expect to be most effective at removing the stain?

   non-polar (Alcohol)

4) If a liquid with a density of 1.2 g/mL is mixed in equal proportions with a liquid that has a density of 0.9 g/mL, what would you expect of the combined density if

   a) the two liquids are completely miscible with additive volumes? *H_2O, Alcohol*
   get sep eventually   no Δ

   b) the two liquids are completely miscible without additive volumes?

   sep

   c) the two liquids are completely immiscible?

   layers

   d) the two liquids are miscible but of greatly varying viscosities?

   layers   lava lamp

SOLUBILITY
REPORT SHEET

Name: _____

Lab Partner: _____

Section: _____ Date: _____

1. DENSITIES OF LIQUIDS

	Syrup	Water	Oil
Mass of cylinder and liquid	_____ g	_____ g	_____ g
Minus mass of cylinder	−_____ g	−_____ g	−_____ g
Mass of liquid	_____ g	_____ g	_____ g

Density of liquids = mass of liquid
                        10 mL

= _____ g/mL   _____ g/mL   _____ g/mL

2. Which liquid is the most dense?

3. Which liquid is on the bottom of the beaker?

4. Which two layers mix in step 4 of the procedure?

	Top Layer	Bottom Layer
Mass of cylinder and liquid	_____ g	_____ g
Minus mass of cylinder	−_____ g	−_____ g
Mass of liquid	_____ g	_____ g

Density of liquids = mass of liquid
                        5 mL

= _____ g/mL   _____ g/mL

no

5. Is the density of the mixture an average of the densities of the original two layers?

Does soap act as an emulsifying agent?

6. Why would forming an emulsion be important when washing greasy dishes?

II. MIXTURE VOLUMES

Volume of water	5 mL
Volume of alcohol	5 mL
	10 mL

Measured volume _____ mL

If the volumes are different, explain why.

III. QUESTIONS

1. Which ingredient in peanut butter allows it to remove gum?

   OIL

2. Why would using peanut butter rather than scissors be better for removing gum from hair?

   choose

3. Why would using peanut butter rather than water be better for removing gum from hair?

   like disslike

4. Would vegetable oil work in a similar manner to peanut butter in removing gum?

5. A substance called lecithin allows vinegar and oil in some dressings to remain mixed. What qualities would you expect lecithin to have in common with soap?

15 *Viscosity*

Thick as Maple Syrup

OBJECTIVES

To gain experience with the property of viscosity and develop an understanding of the
measurement of the property.

To produce solutions of varying viscosities and thereby create a viscosity scale.

To measure the viscosity of various common substances and rate them according to the scale.

To explore the concept of shape and aerodynamic design as related to objects moving
through fluids.

Relates to Chapter 6 of Chemistry for Changing Times, *twelfth ed.*

BACKGROUND

When you tip the ketchup bottle and anticipate the thick, red substance that will flavor your
hamburger, you are witnessing high *viscosity*. Viscosity is a physical property that can be
measured by finding the force required for an object to pass through a fluid.

Viscosity is defined as the resistance of a substance to flow. Several factors are responsible
for the different viscosities we observe in everyday substances. The "thickness" of a fluid is
generally associated with viscosity. Although it is often true, *density* and viscosity are not always
related. The most important factor affecting viscosity is the strength of the cohesive forces
between the particles in the fluid. *Cohesion* is the attraction of one particle for another, identical
particle. The polarity, shape, and size of molecules all play a part in determining the strength of
the cohesive forces between them. If the force of cohesion is greater than the attraction of the
molecules for an object of unlike substance—called *adhesion*—then the fluid will resist
movement of the object through it. A force is necessary to cause the object to move through the
fluid.

In the case of substances containing protein chains, such as ketchup and mustard, viscosity is
the result of a different phenomenon. The proteins attract and then attach to one another as they
sit motionless. Over time, these attachments become quite strong and can even exclude water
molecules, causing a watery film to form on the surface. These substances need to be shaken or
jarred to break up the protein lattices and flow. Hence, many have experienced the familiar
occurrence of slapping the end of the ketchup bottle and then suddenly getting a big burp of
ketchup from the opening.

Since weight is a force, gravity can be used to measure viscosity. Glass marbles are denser
than most fluids and therefore will sink as a result of their weight. If a marble is dropped into a
fluid, the speed at which it falls through the fluid is a measurement of the viscosity. A scale can
be made for comparing viscosities by measuring the time required for a marble to fall through a
specified distance of fluid and changing the viscosity of the fluid by specific intervals.

In this investigation, you will prepare a series of solutions with increasing viscosity and
measure the time required for a marble to sink to the bottom of each, thereby creating a viscosity
scale. Several common substances will then be tested and given a viscosity rating based upon
this scale.

WASTE AND THE ENVIRONMENT

All substances are nontoxic and can be washed down the drain with plenty of water.

⚠ *CAUTION*s warn about safety hazards.
*EXTRA*s give helpful hints, additional information, or interesting facts.

Common Materials
    soluble starch
    vegetable oil
    rubbing alcohol
    honey
    dishwashing liquid
    ketchup

Special Equipment
    glass marbles
    stop watch
    test tube rack
    11 identical test tubes, large
  1 stopper to fit the large test tubes

PROCEDURE
Part A: Producing a Viscosity Scale
1. Weigh one of the test tubes. Record this in the data table.

2. Place 30.0 g of soluble starch into 250 mL of cold water. Stir it vigorously to get the powder evenly dispersed in the water. Warm the water, stirring constantly until it comes to a boil. The starch solution will thicken. When it is cool, pour 30.0 mL of starch solution into one of the test tubes.

3. Place 3 mL water and then 27 mL of the starch solution into a second tube. Stopper and shake until it is homogenous.

4. Place 6 mL water and 24 mL starch solution into the third tube, stopper, and shake. Prepare the remaining tubes with the following water/starch solution ratios: 9:21, 12:18, 15:15, 18:12, 21:9, 24:6, 27:3, 30:0.

5. Take a mass measurement on each of the eleven test tubes. Record these values in the data table.

6. Place a marble into the first tube and stopper it tightly. Invert the tub and hold it until the marble is resting on the stopper. Quickly turn the tube upright and start the stopwatch at the same time. Measure the number of seconds required for the marble to reach the bottom of the tube. Record this value in the data table. Remove, rinse, and dry the marble and the stopper.

7. Repeat this for all eleven test tubes. It may be necessary in the interest of time to note the time on the clock for the first two tubes and continue making measurements on the others as you wait for the marble to settle to the bottom.

Part B: Determination of Relative Viscosity from Scale
8. Add 30.0 mL of each of the following substances to five separate test tubes: vegetable oil, rubbing alcohol, honey, dish-washing liquid, and ketchup.

9. Make a mass measurement for each of the new substances. Making several measurements for each tube and then recording the average may yield better results.

10. Repeat the time measurement as directed in step 6 for each of the five new substances, and record the values in the data table.

11. Rinse all substances down the drain with plenty of water. Clean and dry all test tubes and stoppers. Making sure the marble and the stopper are clean and dry from the previous solution eliminates the risk of contamination that could alter the measurement.

EXTRA
Be sure to remove all lumps of powder before beginning to heat the mixture.

EXTRA
Making all the volumes the same eliminates one of the variables that could affect the time measurement: the depth of the fluid.

EXTRA
It may help to set the tubes in a small beaker to take the masses.
Don't forget to subtract the mass of the beaker.

EXTRA
Starting all the marbles on the stopper helps eliminate another variable for the time measurement: the distance it drops.

VISCOSITY
PRE-LAB QUESTIONS

Name: _____

Lab Partner: _____

Section: _____   Date: _____

1) Starch is a linear protein molecule. Knowing this, what might you expect to be the effect on the viscosity of a starch solution if it was allowed to sit for a while?

2) If two objects of the same material, density, and mass, but of different shape, are each allowed to fall through the same viscous fluid, how would shape affect the apparent measurement of the fluid's viscosity?

3) Considering the cohesive forces between particles in a highly volatile fluid—that is, one that evaporates easily—what would you suspect concerning the viscosity of the fluid? Explain your reasoning.

4) Air is not considered to be a very viscous fluid. However, automobiles and planes are designed with aerodynamics in mind. How does the speed of an object in a fluid affect the resistance force it experiences from the viscosity?

VISCOSITY
REPORT SHEET

Name: _____

Lab Partner: _____

Section: _____   Date: _____

I. PART A: PRODUCING A VISCOSITY SCALE
Data Table

Mass of one test tube:_____ g   Volume in each tube: 30.0 mL

Test Tube	Water/Starch Ratio	Mass	Time for Marble to Reach Bottom
1	0:30	_____ g	_____ seconds
2	3:27	_____ g	_____ seconds
3	6:24	_____ g	_____ seconds
4	9:21	_____ g	_____ seconds
5	12:18	_____ g	_____ seconds
6	15:15	_____ g	_____ seconds
7	18:12	_____ g	_____ seconds
8	21:9	_____ g	_____ seconds
9	24:6	_____ g	_____ seconds
10	27:3	_____ g	_____ seconds
11	30:0	_____ g	_____ seconds

Calculations Table
Density Calculation

Test Tube	Total Mass	– Test-Tube Mass	= Mass of Solution	Density = Mass/Volume
1	_____ g	– _____ g	= _____ g	_____ g/mL
2	_____ g	– _____ g	= _____ g	_____ g/mL
3	_____ g	– _____ g	= _____ g	_____ g/mL
4	_____ g	– _____ g	= _____ g	_____ g/mL
5	_____ g	– _____ g	= _____ g	_____ g/mL

6 _____ g − _____ g = _____ g _____ g/mL

7 _____ g − _____ g = _____ g _____ g/mL

8 _____ g − _____ g = _____ g _____ g/mL

9 _____ g − _____ g = _____ g _____ g/mL

10 _____ g − _____ g = _____ g _____ g/mL

11 _____ g − _____ g = _____ g _____ g/mL

Viscosity Scale
Divide each time measurement by the smallest one to get a ratio. Give the ratio to three significant digits.

Test Tube   Viscosity Scale

1 _____

2 _____

3 _____

4 _____

5 _____

6 _____

7 _____

8 _____

9 _____

10 _____

11 _____

II. Part B: DETERMINATION OF RELATIVE VISCOSITIES OF COMMON SUBSTANCES
Data Table

Substance	Time to Bottom of Tube	Mass Measurement	Relative Viscosity from Scale
Vegetable pil	_____ seconds	_____ g	_____
Rubbing alcohol	_____ seconds	_____ g	_____
Honey	_____ seconds	_____ g	_____
Dish liquid	_____ seconds	_____ g	_____
Ketchup	_____ seconds	_____ g	

Calculations Table

Substance	Mass	– Test-Tube Mass	= Mass of Substance	Density = Mass/Volume
Vegetable oil	_____ g	– _____ g =	_____ g	_____ g/mL
Rubbing alcohol	_____ g	– _____ g =	_____ g	_____ g/mL
Honey	_____ g	– _____ g =	_____ g	_____ g/mL
Dish liquid	_____ g	– _____ g =	_____ g	_____ g/mL
Ketchup	_____ g	– _____ g =	_____ g	_____ g/mL

III. QUESTIONS
1. Do the densities of the starch solutions vary directly with the viscosity ratings?

2. Do the densities of the other substances vary directly with the viscosity ratings?

3. If you answered differently for the first two questions, give an explanation for the apparent discrepancy.

4. What factors, other than the viscosity of the fluid, affect how fast the marble falls down the test tube? Does this in any way affect the results as they have been gathered? Why or why not?

5. How would the data be affected by using a pointed object to fall through the fluids rather than a spherical one?

125

6. Relate your answer in question 5 to the general shape of fish and submarines. How is this applicable to automobiles that also travel through a gaseous fluid, air?

7. Honey and molasses are both sugar solutions. Knowing this, and that molasses is more dense than honey, can you make a prediction as to the viscosity of molasses? Why is this generalization okay to use when comparing these two substances?

16 *Gas Laws*

It's Not Just Hot Air

OBJECTIVES

To learn how temperature, pressure, and volume of a gas are related.

To determine how temperature, pressure, and volume in a gas affect each other.

To calculate the ideal gas law constant, "R."

To measure the amount-volume relationship by holding other variables constant.

Relates to Chapter 6 of Chemistry for Changing Times, *twelfth ed.*

BACKGROUND

A minimum of four variables is required to completely describe a gas—the *amount* measured in moles, *volume*, *temperature* measured in the absolute Kelvin scale, and *pressure*. Each of the four quantities is dependent upon the other three. Gas behavior is described by a series of mathematical laws that have been discovered and experimentally proven by scientists. Each of these laws describes the relationship between two of the following variables: pressure, temperature, volume, and number of moles.

Boyle's law states that for a given amount of gas at a constant temperature, the product of the pressure and volume is a constant. Another way to write this is $PV = a$, where a is a constant. As the pressure increases, the volume decreases. As a weather balloon rises into the higher atmosphere where the pressure is lower, the balloon swells. The first pressure measurements were made using an open-ended manometer similar to the apparatus used in this investigation (procedure 1) except that mercury was used instead of water. Mercury is 13.6 times denser than water, therefore the difference in the fluid heights should be a factor of 13.6 more in our investigation. Pressure is still sometimes reported in the pressure unit "millimeters of mercury," or mmHg. (1 mmHg is the same as 1 torr. One "atmosphere" of pressure is equal to 760 mmHg or 760 torr.)

Charles's law states that the volume of a fixed amount of a gas at a constant pressure is directly proportional to its Kelvin (K) temperature. Another way to write this relationship is $V = bT$ or $\frac{V}{T} = b$, where b is a constant. As the temperature decreases, the volume decreases. When volume is plotted versus the Kelvin temperature, the straight line generated can be extrapolated to find the temperature at zero volume. Scientists have done this experimentally and the temperature was found to be -273 °C. This temperature is 0 on the Kelvin scale and is called *absolute zero*.

The relationship between pressure and temperature is sometimes called *Amontons' law*. It is also called *Gay-Lussac's law*. It states that the pressure of a fixed amount of gas at a constant volume is proportional to its Kelvin temperature. This law can also be written as $P = cT$ or $\frac{P}{T} = c$, where c is a constant. On a long automobile trip, the pressure of the air in the tires increases because the flexing of the tires causes an increase in the temperature.

Avogadro's law states that a fixed volume of any gas at a constant temperature and pressure contains the same number of molecules. It is now known that the same number of molecules means the same number of moles of gas. The law can also be written as $V = dn$ or $\dfrac{V}{n} = d$, where d is a constant. For example, a balloon is inflated by adding molecules of gas to the interior.

All the gas laws can be combined to produce the *ideal gas law*, $\dfrac{PV}{nT}$ = constant. The constant, "R," is called the universal gas constant and is equal to $\dfrac{0.08206\,\text{L}\cdot\text{atm}}{\text{K}\cdot\text{mol}}$. The ideal gas law works well for most gases at normal temperatures and pressures. Very precise measurements will show that each gas acts a little differently and that real gases are non-ideal gases especially at low temperatures and high pressures.

WASTE AND THE ENVIRONMENT
The compounds used in this investigation are not toxic.

⚠ *CAUTION*s warn about safety hazards.
*EXTRA*s give helpful hints, additional information, or interesting facts.

Reagents
    solid sodium hydrogen carbonate ($NaHCO_3$)
    1 M hydrochloric acid
Laboratory Equipment
    very small (75 mm long) test tube
    clear tubing, about 4 ft.
    100 mL graduated cylinder and 1-hole stopper
    250 mL Erlenmeyer flask and 1-hole stopper
    thermometer

Common Materials
    drinking straw
    ruler (cm)

    #1 1-hole rubber stopper with short glass tube
    10 mL graduated cylinder
    buret
    1000 mL beaker
    barometer

PROCEDURE
Boyle's Law
1. Attach a 4-ft length of clear tubing to the tip of a buret. Attach the other end of the clear tubing to a piece of glass tubing fit through a #1 rubber stopper. Half fill a 10 mL graduated cylinder with water. Fill the buret two-thirds full. Open the stopcock. Hold the buret high enough so that water streams from the stopper with no bubbles. Tip the graduated cylinder to quickly insert the stopper into the cylinder. See Figure 1.

   EXTRA
   It is very important to avoid bubbles in the tube. Bubbles cause a large error in the results.

2. Read the barometric pressure on a barometer and record it in mmHg.

3. With the graduated cylinder upside down, hold the buret beside the cylinder so that the top of the water in each is at the same level. Read the volume of air in the graduated cylinder.

   EXTRA
   Remember that the graduated cylinder readings are upside down.

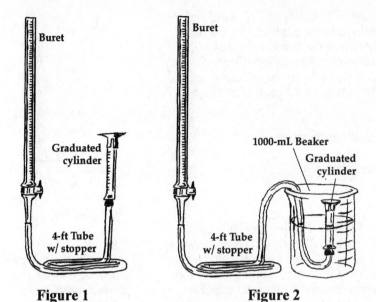

Figure 1   **Figure 2**

4. It will be obvious that as the buret is lifted it causes an increase in the pressure on the gas in the cylinder and the volume of the gas decreases. Lift the buret to a certain point. Read the volume of gas. Measure the distance in millimeters (mm) from the water level in the graduated cylinder to the water level in the buret. Record these measurements. Lift the buret to another height. Record air volume and water-height difference.

5. Lower the buret to a point where the buret water level is below the cylinder water level. This reduces the pressure and will increase the volume. Measure and record the volume and pressure for two different reduced pressures.

Charles's Law
6. Using the same setup, allow the cylinder to sit for a few minutes to reach room temperature. Raise or lower the cylinder to make the two water levels even. Read and record the volume of air and the room temperature.

7. Immerse the cylinder (base up) in a beaker of very cold water. See Figure 2. Allow it to sit for several minutes. Raise or lower the cylinder to make the two water levels even. Read and record the volume of air and the temperature (K) of the water.

8. Repeat step 7 with a beaker of warm water.

9. Repeat step 7 with a beaker of hot water.

Gay-Lussac's Law
10. Using the same setup, allow the cylinder to sit for a few minutes to reach room temperature. Raise or lower the cylinder to make the two water levels even. Read and record the volume of air, the room temperature (K), and the barometric pressure.

EXTRA
Divide the distance between water heights by 13.6 (the density of mercury) and add it to the barometric pressure to get the pressure of the gas in the cylinder.

EXTRA
Divide the distance between water heights by 13.6 (the density of mercury) and subtract it from the barometric pressure to get the pressure of the gas in the cylinder.

⚠ *CAUTION*
Do not burn yourself. Hold the cylinder with tongs or a test-tube holder.

11. Immerse the cylinder (base up) in a beaker of very cold water. See Figure 2. Allow it to sit for several minutes. Adjust the cylinder so that the volume of air is the same as in step 10. Measure the water height difference. Divide the distance between water heights by 13.6 (the density of mercury) and subtract it from the barometric pressure to get the pressure of the gas in the cylinder.

12. Repeat step 11 with a beaker of warm water.

13. Repeat step 11 with a beaker of hot water.

Ideal Gas Law

14. Using a straw, blow into the water in a filled 1000 mL beaker so the water is saturated with carbon dioxide (CO_2). Pour water from the beaker to fill a 100 mL graduated cylinder. Place a stopper connected by a glass tube to flexible tubing in the top of the cylinder. Tip the cylinder into the beaker. Clamp the cylinder in place. If any air got into the cylinder, read the water level. Attach a stopper to the tubing to seal a 250 mL Erlenmeyer flask.

⚠ CAUTION
Do not burn yourself. Hold the cylinder with tongs or a test-tube holder.

EXTRA
Be sure the stopper does not block the pour spout of the cylinder and prevent the water from leaving as the gas is collected.

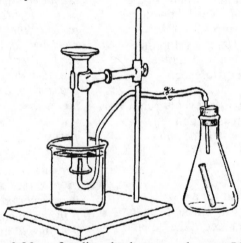

15. Place 0.20 g of sodium hydrogen carbonate ($NaHCO_3$) in the flask. Place 3 mL of 1 M hydrochloric acid (HCl) in a 75-mm long test tube. Place the test tube in the flask. Securely place the stopper in the flask. Tip the flask so that the acid reacts with the sodium carbonate. Allow all the gas to collect in the cylinder.

EXTRA
$NaHCO_3$ is also called sodium bicarbonate.

16. Record the temperature (K) of the water and the barometric pressure. Unclamp the cylinder. Adjust the height of the cylinder so that the water levels are even. Read and record the gas volume.

EXTRA
1 atm is equal to 760 mmHg.

17. You may pour the solutions down the drain.

GAS LAWS
PRE-LAB QUESTIONS

Name: _____

Lab Partner: _____

Section: _____   Date: _____

1) In step 1, air is trapped inside the small cylinder above the water level. The air in the top of the buret is open to the atmosphere. As the buret is raised and lowered, the water exerts varying pressures on the trapped gas in the small cylinder. How is the gas pressure at the top of the water in the buret affected? Explain.

2) Gas volumes change with temperature when all other factors are held constant. What would you expect to see happen to a fully inflated Mylar® balloon if it was taken outside into freezing weather? Why?

3) Basketballs inflated at one temperature and then used at another may not behave as expected. Why do balls often seem "flat" in the wintertime?

4) If a gas is collected in a rigid container so that volume, temperature, and pressure are known, what relationship will allow the experimenter to find the amount of gas collected?

GAS LAWS
REPORT SHEET

Name: _____

Lab Partner: _____

Section: _____   Date: _____

I. BOYLE'S LAW

	Air Volume	Pressure	PV
Even levels	_____	_____	_____
First raised buret	_____	_____	_____
Second raised buret	_____	_____	_____
First lowered buret	_____	_____	_____
Second lowered buret	_____	_____	_____

II. CHARLES'S LAW

	Temperature (K)	Volume	$\dfrac{V}{T}$
Room temperature	_____	_____	_____
Lower temperature	_____	_____	_____
Higher temperature	_____	_____	_____
Highest temperature	_____	_____	_____

III. GAY-LUSSAC'S LAW

Volume _____

	Pressure	Temperature (K)	$\dfrac{P}{T}$
Room temperature	_____	_____	_____
Lower temperature	_____	_____	_____
Higher temperature	_____	_____	_____
Highest temperature	_____	_____	_____

IV. IDEAL GAS LAW

Final volume − Initial volume   Gas volume

_____mL − _____mL = _____mL $\times \dfrac{10^{-3}\,L}{mL}$ = _____ **L(V)**

Temperature _____°C +<u>273</u>° = _____**K (T)**

Pressure _____mmHg $\times \dfrac{1\,atm}{760\,mmHg}$ = _____**atm (P)**

Amount _____g $NaHCO_3 \times \dfrac{1\,mol}{84\,g}$ = _____**mol (n)**

$R = \dfrac{PV}{nT}$ = _____ $\dfrac{L \cdot atm}{K \cdot mol}$

V. QUESTIONS

1. Are the values for PV all the same?

2. Are the values for $\dfrac{V}{T}$ all the same?

3. Are the values for $\dfrac{P}{T}$ all the same?

4. What was the percent error of your value for R?

$$\dfrac{R_{actual} - R_{calculated}}{R_{actual}} \quad \times \quad 100\% \quad = \quad \% \text{ error}$$

5. If you plotted your temperature-volume data, at what temperature would the volume be zero?

6. If you check the tire pressure after the car has been driven long enough for the tires to gain heat, what would you find?

17 *Acid Neutralization by Antacid*

How to Stop Heartburn

OBJECTIVES

To achieve neutralization in an acid-base titration reaction.

To perform titration by the gravimetric method.

To observe the use of the indicator phenolphthalein in a titration.

To calculate the amount of acid absorbed per gram of antacid based on data gathered from titrations.

Relates to Chapter 7 of Chemistry for Changing Times, *twelfth ed.*

BACKGROUND

Stomach acid is a combination of gastric juices and an acid very similar to hydrochloric acid. Sometimes eating rich food or experiencing stress may cause more than the usual amount of stomach acid to be produced, causing discomfort.

Commercial *antacids* are primarily composed of basic, or alkaline, compounds and binders to hold the tablet together. Sometimes fillers are also added. The basic compounds react with, or neutralize, the excess stomach acid that causes "acid stomach" or "heartburn." Some of the basic compounds used are *hydroxides* such as magnesium hydroxide, $Mg(OH)_2$, or aluminum hydroxide, $Al(OH)_3$. The hydroxide ion is released when the hydroxide dissolves in water. The reaction between the hydroxide ion and the acidic hydrogen ion reduces the amount of acid and relieves the discomfort.

$$Mg(OH)_2(s) \xrightarrow{H_2O} Mg^{2+} + 2\,OH^-$$

$$OH^- + H^+ \rightarrow H_2O$$

The free hydrogen ion does not actually exist. It is always combined with some molecule. In an aqueous solution, the hydrogen ion combines with water and forms ions such as H_3O^+, called *hydronium*, or $H_5O_2^+$. Because we do not know the exact form of the ion, it is easier just to write the hydrogen ion as H^+ or $H^+(aq)$.

Other basic compounds used as antacids are *carbonates* such as calcium carbonate ($CaCO_3$), magnesium carbonate ($MgCO_3$), and sodium carbonate (Na_2CO_3). These react with the hydrogen ion to form carbonic acid, which quickly dissociates into water and carbon dioxide. Carbon dioxide is a gas and may cause belching, which in turn also helps relieve stomach distress.

$$CO_3^{2-} + 2\,H^+ \rightarrow H_2CO_3 \rightarrow H_2O + CO_2$$

This investigation involves doing a *back titration*. A back titration is the titration of an excess of reagent. An antacid will be added to a solution that simulates stomach acid. Enough antacid will be added to leave an excess amount of acid in the solution. The excess acid is the acid that was not neutralized by the antacid. The amount of base necessary to neutralize the remaining acid in the solution is equal to the amount of excess acid. Thus, the less base required to reach neutrality, the more acid that was absorbed by the antacid. The reaction is

$$\text{from acid} \longrightarrow H^+ \; + \; OH^- \longleftarrow \text{from base} \rightarrow H_2O$$

Phenolphthalein is a colorless compound in its acid form. When it loses a hydrogen ion to become the base form, it becomes pink. Because phenolphthalein changes color when it loses its hydrogen ion to bases, it is used as an *indicator*.

Which antacid neutralizes the most acid? It is time for you to find out.

WASTE AND THE ENVIRONMENT
Concentrated acids and bases can damage plumbing if not neutralized or diluted. All final solutions in the lab will be neutral and can be flushed down the drain with plenty of water.

⚠ *CAUTION*s warn about safety hazards.
*EXTRA*s give helpful hints, additional information, or interesting facts.

Reagents
6-M hydrochloric acid [HCl]   distilled water
sodium hydroxide pellets [NaOH]   phenolphthalein

Common Materials
antacid tablets, 3 different brands (Alka-Seltzer® suggested)

Laboratory Equipment
1 50 mL pipette with safety bulb   100 mL clean squeeze bottle
1 500 mL volumetric flask   500 mL Erlenmeyer flask
solid stopper to fit volumetric flask   solid stopper to fit Erlenmeyer flask
1 50 mL graduated cylinder   150 mL Erlenmeyer flask
mortar and pestle   ring, stand, and wire gauze
centigram balance   laboratory burner or hotplate

PROCEDURE

1. Rinse a 500 mL volumetric flask well with water and then add about 400 mL of distilled water to the flask. Using a graduated cylinder, add exactly 30 mL of 6.0 M hydrochloric acid (HCl). Fill the flask almost to 500 mL with distilled water. Stopper the flask and invert it several times to mix the solution well. Add the last amount of water one drop at a time. The flask contains exactly 500 mL when the bottom of the meniscus is level with the etched line on the neck of the flask. (See page 486 for help on reading a meniscus.) The resulting solution is 0.36 M hydrochloric acid.

▲ *CAUTION*
Hydrochloric acid is corrosive.

2. To prepare 1 M sodium hydroxide (NaOH), weigh a small, clean beaker. Add about 20 g of sodium hydroxide pellets to the beaker. Remove the beaker from the balance. Add 100 mL of distilled water to dissolve the pellets. Pour the sodium hydroxide solution, including any undissolved pellets, into a clean 500 mL Erlenmeyer flask. Rinse the beaker into the Erlenmeyer flask with small portions of distilled water. Fill the Erlenmeyer flask to about 500 mL. Mix to be sure all the pellets are dissolved. Keep the solution stoppered.

▲ *CAUTION*
Sodium hydroxide is caustic.

▲ *CAUTION*
Dissolving sodium hydroxide generates heat. Do not burn yourself.

3. **Density of NaOH Solution**
Weigh a clean, dry 10 mL graduated cylinder. Record the mass. Fill the cylinder with exactly 10 mL of sodium hydroxide solution. Reweigh the cylinder containing the solution and record the new mass. The 10 mL of solution can be poured back into your beaker of sodium hydroxide to be used in step 5.

4. **Standardization of the Sodium Hydroxide Solution**
Using a safety bulb, pipet 50 mL of 0.36 M hydrochloric acid solution into a 150 mL Erlenmeyer flask. Add 2 or 3 drops of phenolphthalein.

EXTRA
Sodium hydroxide solution cannot be made in a known concentration, because it absorbs moisture from the air and dissolves. It is deliquescent.

5. Pour about 50 mL of sodium hydroxide solution into a clean, dry squeeze bottle. Weigh the bottle and solution and record the weight as the initial weight. The next step is to "titrate" the acid solution with the sodium hydroxide solution to determine the amount of acid (hydrogen ions) in solution. To titrate, add the sodium hydroxide slowly until the "endpoint." At the endpoint, the solution will stay pink for at least 30 seconds. At the beginning of the titration, you add a large squirt. Note the pink color that fades with swirling. As the pink remains longer, add smaller portions. Add the last amount one drop at a time. Stop adding when one drop causes a pink color that does not fade within 30 seconds. Reweigh the bottle and record the mass as the final mass.

EXTRA
Phenolphthalein turns pink in a basic solution.

6. **Neutralization with Antacids**
Weigh two antacid tablets of the same brand. Record the brand and the mass. Use a mortar and pestle to powder the tablets.

7. Dissolve the tablets in 50 mL of 0.36 M hydrochloric acid. This is an excess of acid. Add 2 or 3 drops of phenolphthalein. Slightly warm the solution over a laboratory burner or hotplate if necessary to completely dissolve the tablets.

EXTRA
Stomach acid is similar to 0.36 M hydrochloric acid.

8. Refill the squeeze bottle with about 50 mL of sodium hydroxide solution. Weigh the bottle and record the mass as the initial mass.

9. Using the squeeze bottle, titrate the excess acid in the tablet solution to a faint pink endpoint, just as in step 5. Reweigh the bottle and record the mass as the final mass.

10. Repeat steps 6 through 9 for two more brands of antacid tablets. To test Alka-Seltzer, dissolve 2 tablets in 50 mL of distilled water. Add 50 mL of 0.36-M hydrochloric acid and 3 drops of phenolphthalein. Titrate with the squeeze bottle of sodium hydroxide.

EXTRA
My stomach feels better now.

11. Flush all solutions down the drain with a lot of water unless otherwise directed by your instructor.

ACID NEUTRALIZATION BY ANTACID
PRE-LAB QUESTIONS

Name: _____

Lab Partner: _____

Section: _____   Date: _____

1) Some antacids neutralize acid very quickly while others neutralize acid slowly over a longer period of time. The antacids that produce carbon dioxide gas neutralize acid the quickest while antacids containing hydroxides tend to operate over a longer period of time. Name a common antacid that offers immediate relief.

2) If an antacid tablet requires warming to dissolve, or if it dissolves slowly, would you expect it to work quickly or be designed to operate over an extended period of time?

3) What color would you expect phenolphthalein to be if it was added to the gastric juices of your stomach?

4) Why would it not be a good idea to use antacids as a calcium supplement?

5) What happens to the pH of your stomach if you take more antacid than necessary to neutralize the acid?

ACID NEUTRALIZATION BY ANTACID
REPORT SHEET

Name: _____

Lab Partner: _____

Section: _____   Date: _____

I. DENSITY OF NaOH

Mass of cylinder and NaOH solution   _____ g

Minus mass of cylinder   –   _____ g

Mass of 10 mL of NaOH solution   _____ g

$$\text{Density} = \frac{\text{mass of solution}}{\text{volume}} = \frac{\text{_____ g}}{10\,\text{mL}} = \underline{\hspace{2cm}} \text{ g/mL}$$

II. STANDARDIZATION OF NaOH

Mass of bottle

    Initial mass   _____ g

    Minus final mass   –   _____ g

    Mass of NaOH added   _____ g

$$\text{Volume of NaOH added} = \frac{\text{mass added}}{\text{density}} = \frac{\text{_____ g}}{\text{_____ g/mL}} = \underline{\hspace{1.5cm}} \text{ mL}$$

$$\text{Molarity of NaOH} = \frac{(0.36\text{ M HCl})(50\text{ mL HCl})}{\text{volume NaOH}}$$

$$= \frac{(0.36\text{ M HCl})(50\text{ mL HCl})}{\underline{\hspace{1.5cm}}\text{mL NaOH}} = \underline{\hspace{3cm}} \text{M}$$

	#1	#2	#3
III. BRAND NAME OF TABLETS	_____	_____	_____
Mass	_____ g	_____ g	_____ g
IV. BASE ADDED Initial solution and bottle mass	_____ g	_____ g	_____ g
Final solution and bottle mass	_____ g	_____ g	_____ g

Chemical Investigations for Changing Times

Difference   _____g   _____g   _____g
(Base added)

Volume of base
= difference
density   _____mL   _____mL   _____mL

V. ACID ADDED =
    (50 mL)(0.36 M) =   18 mmol   18 mmol   18 mmol

VI. BASE ADDED =
    volume times
    molarity   = _____ mmol _____mmol   _____mmol

VII. ACID NEUTRALIZED =
    Acid added
    minus base added = _____ mmol _____mmol   _____mmol

VIII. ACID NEUTRALIZED
     PER TABLET   = _____mmol _____mmol_____mmol

IX. ACID NEUTRALIZED
    PER GRAM =
    (answer to VIII divided
    by mass of tablet) = _____mmol _____mmol_____mmol

X. QUESTIONS

1. Read the antacid label to find the basic compounds and write
   balanced equations showing the reaction with the hydrogen ion.

   Brand 1

   Brand 2

   Brand 3

2. Which antacid appears to be best at relieving acid stomach?

3. Which antacid has more neutralizing power per gram?

4. In recent commercials, some antacids brag that they contain calcium. If calcium is used
   by the body to form bones and teeth, is it an advantage to be able to obtain calcium for
   the body as heartburn is treated?

18 *Nature's Indicators*

Don't Fool Mother Nature

OBJECTIVES

To understand the function of an indicator.

To grasp the difference between single transition indicator and a universal indicator.

To produce an indicator from a common household source.

To produce color standards for the common household source indicator.

To estimate the pH of several common substances based on the color produced by the indicator.

Relates to Chapter 7 of Chemistry for Changing Times, *twelfth ed.*

BACKGROUND

There are many compounds that change color as the acidity of a solution changes. We use some of these as acid-base *indicators* because they change color at a certain acidity, or H^+ concentration. Some indicators produce a variety of colors over a wide range of H^+ concentrations. These *universal* indicators are combinations of several indicators that each undergo a specific color change over a certain range of acidity.

The concentration of hydrogen ions can be measured in *molarity* (M), or moles of H^+ per liter of solution, like most other ions in solution. Because the H^+ concentration varies over a wide range, chemists use a convenient scale called pH. The pH is the negative log of the molar concentration of the hydrogen ion. Thus, pH is the measure of acidity or basicity. A neutral solution has a pH of 7. pH values of less than 7 indicate an acidic solution, with each unit change denoting a tenfold change in acidity. Values of pH greater than 7 indicate a basic solution.

pH	H^+ concentration (M)	Interpretation
1	10^{-1}	Very acidic
4	10^{-4}	Acidic
7	10^{-7}	Neutral
10	10^{-10}	Basic
13	10^{-13}	Very basic

Many plants contain chemicals that change color in response to a change of acidity. Possibly the best known of these indicators is *litmus*, which is blue in basic solution and red in acidic solution. Litmus is actually a combination of several compounds. It is chiefly azolitmin and erythrolitmin combined with alkalis such as lecanoric acid, orcein, and erythrolein. Azolitmin, erythrolitmin, and orcein are also combinations of several compounds.

Purple cabbage and radishes contain a class of chemical compounds called *anthocyanins*, which produce a more varied reaction to acidity change than does litmus. The juices extracted from black grape skins or beets, as well as concentrated grape juice, also function as indicators that give evidence of varied hydrogen ion concentrations. As shown in the following structures, the anthocyanins have several OH groups attached to *aromatic* groups. Aromatic groups are ring structures made of carbon atoms and benzene is an example of an aromatic compound. The OH groups are slightly acidic. As the H^+ concentration decreases, more H^+ ions are removed from the cyanin structure. As the number of H^+ ions change, the wavelength of light absorbed by the molecule changes, so the color of the solution changes. Thus, anthocyanins can act as acid-base indicators. The anthocyanin cyanidine occurs in two common flowers. In the poppy, it is in the acid form, so the flower is red. The cornflower is blue because the cyanidine is in the base form. Anthocyanins such as pelargonidin and delphinidin are the source of color for flowers in the genera and species *Pelargonium zonale* and *Delphinium consolida L.*

Red   Cyanidine   Blue

Pelargonidin   Delphinidin

We will test several household substances to find their pH by using a reference color scale derived from a natural universal indicator. Lemon juice contains citric and ascorbic acids. Vinegar contains acetic acid, and ammonia is a base. Soft drinks are often acidified to keep them from being too sweet. Acidic solutions are sour to the taste. Carbon dioxide is dissolved in water to form carbonic acid or carbonated water. This makes the drink acidic. As the carbon dioxide is lost (the drink goes flat) the drink becomes sweeter, and thus less acidic.

Drain cleaners for the kitchen are basic. Bases react with the grease in kitchen clogs to form soap, which helps free the clog. Bathroom drain cleaners contain acids to break down proteins such as hair.

WASTE AND THE ENVIRONMENT
Concentrated acids and bases can damage plumbing if not neutralized or diluted.

⚠ *CAUTION*s warn about safety hazards.
*EXTRA*s give helpful hints, additional information, or interesting facts.

Reagents
    0.1 M hydrochloric acid [HCl]   pH test paper
    0.1 M sodium hydroxide [NaOH]   red and blue litmus paper
Common Materials
    red cabbage   vinegar
    beets   clear soft drink
    conc. grape juice   white or clear shampoo (clear is better)
    lemon juice   liquid drain cleaner
    ammonia household cleanser   knives
Laboratory Equipment
    250 mL beakers   ring stand, ring, and wire gauze
    7 test tubes   eyedropper
    100 mL graduated cylinder   laboratory burner
    7 100 mL beakers   stirring rod

Assign set to one group   x Δ Results

PROCEDURE

1. Chop one-fourth of a small red cabbage into small pieces. Place the pieces in a 250 mL beaker with 100 mL of water. Boil gently until the color turns dark. Set the mixture aside to cool. Pour off the cabbage extract, or indicator, into a beaker labeled "red cabbage," leaving the solid behind.

2. Cut a beet into small pieces. Place the pieces in a 250 mL beaker with 100 mL of water. Boil the mixture gently until the water turns a dark color. Set it aside to cool. Pour off the beet extract, or indicator, into a beaker labeled "beet," leaving the solid behind.

3. Obtain about 20 mL of 0.1 M HCl and place it in a beaker labeled pH = 1.

4. Place 1.00 mL of the 0.1 M HCl in a 100 mL graduated cylinder. Fill the cylinder with distilled water to exactly 100 mL. Pour this solution into a beaker labeled pH = 3.

5. Rinse the cylinder several times with water. Transfer 1.00 mL of pH = 3 solution to the cylinder and fill with water to exactly 100 mL. Pour this solution into a beaker labeled pH = 5.

6. Place distilled water that has been boiled then allowed to cool, while covered, in a beaker labeled pH = 7.

7. Obtain about 20 mL of 0.1 M NaOH and place it in a beaker labeled pH = 13.

8. Place 1.00 mL of the 0.1 M NaOH in a clean 100 mL graduated cylinder. Fill the cylinder to exactly 100 mL. Pour this solution into a beaker labeled pH = 11.

9. Rinse the cylinder several times with water. Place 1.00 mL of the pH = 11 solution in the cylinder. Fill the cylinder with water to exactly 100 mL. Pour this solution into a beaker labeled pH = 9.

10. Pour about 1 mL of each of the pH solutions into each of seven test tubes and add several drops of red cabbage indicator to each test tube. Record the color at each pH.

11. Using a clean glass stirring rod, touch the solution in each test tube and then touch one type of litmus paper. Repeat for the other type of litmus paper. Clean and dry the stirring rod between solutions.

12. Rinse the test tubes several times and repeat step 10 using beet juice indicator and then grape juice. Record the results.

All test Substances

⚠ *CAUTION*
Don't burn yourself.

EXTRA
Use only enough water to cover the cabbage.

EXTRA
Canned red cabbage can be used and is quicker.

⚠ *CAUTION*
HCl is caustic.

EXTRA
Normal water is acidic because of absorbed CO_2.

⚠ *CAUTION*
NaOH is caustic.

EXTRA
Litmus paper is used instead of pH paper because litmus is another naturally occurring indicator.

EXTRA
When making a grape juice indicator, use only the skins of black grapes.

13. Choose one indicator and use it to approximate the pH of lemon juice, vinegar, a clear soft drink, a clear shampoo, household ammonia cleanser, and liquid drain cleaner. Test each solution by placing 1 mL of the solution into a rinsed test tube and adding several drops of indicator.

14. Use pH paper to determine the pH of each substance by touching the solution and then the pH paper with a clean glass stirring rod. Be sure to clean and dry the stirring rod between solutions.

15. Dispose of your pH solutions one at a time by pouring them down the drain with a large amount of water, unless otherwise directed by your instructor.

⚠ *CAUTION*
Pouring the acids and bases together can create excess heat and splattering.

NATURE'S INDICATORS
PRE-LAB QUESTIONS

Name: _____

Lab Partner: _____

Section: _____   Date: _____

1) Hydrangea plants (*Hydrangea macrophylla*) produce very large flower clusters and are common garden shrubs in central and southern states. Many people believe that there are two varieties of this shrub—blue and pink—when, in fact, there is only one. Many homeowners have planted a hydrangea of one "variety" only to have it slowly transform to the other over a period of years. Give an explanation for this phenomenon in light of the introductory information for this investigation.

Sunlight - Rxn

2) Litmus is either red or blue. It is useful only to determine if a substance is classified as an acid or a base. Indicators that are capable of specifying a particular pH exhibit a range of color changes. Would you expect that these "universal indicators" are a single compound, two compounds combined, or several compounds combined? Support your position.

Several

3) Would an experiment using indicators to determine the pH of common substances work in an already colored substance, such as a cola beverage? Why or why not?

Touch color effects

4) Nearly all indicators are organic molecules that are sensitive to destruction by strong solutions, especially bleaching cleansers. Could the pH of chlorine bleach be determined with an indicator derived from a plant? Explain.

Touch = Damage plant

5) A child goes to the kitchen sink to rinse the glass from which he just drank grape juice. As the water runs into the glass the juice residue turns from purple to light blue. What is happening? *Dilution*

NATURE'S INDICATORS
REPORT SHEET

Name: _____

Lab Partner: _____

Section: _____ Date: _____

I. INDICATOR COLOR

	Cabbage	Beet	Grape	Blue Litmus	Red Litmus
pH = 1	_____	_____	_____	_____	_____
pH = 3	_____	_____	_____	_____	_____
pH = 5	_____	_____	_____	_____	_____
pH = 7	_____	_____	_____	_____	_____
pH = 9	_____	_____	_____	_____	_____
pH = 11	_____	_____	_____	_____	_____
pH = 13	_____	_____	_____	_____	_____

II. INDICATOR USED _____

SUBSTANCE	Using pH Indicator	Using pH Paper
Lemon juice	_____	_____
Vinegar	_____	_____
Soft drink	_____	_____
Shampoo	_____	_____
Ammonia cleanser	_____	_____
Drain cleaner	_____	_____

III. QUESTIONS

1. Are beverages usually acidic or basic?

2. Why shouldn't shampoo be too acidic or too basic?

3. Which indicator did you like best and why?

4. Which indicator covers the largest range of pH?

5. Which indicator produced the most varying colors?

6. Why are there two types of drain cleaners?

7. Comment on how closely your pH determinations using the experimental indicator scale matched the pH determinations with the commercial pH paper.

19 pH and Flammability

Safety in the Home!

OBJECTIVES

To discover the wide pH range represented by household products.

To gain an appreciation for the flammability of common household products.

To develop a respect for the inherent dangers of many common household products.

Relates to Chapter 7 from Chemistry for Changing Times, *twelfth ed.*

BACKGROUND

The American home contains a large variety of chemical products. There are products to clean clothes, windows, ovens, drains, floors, and people. Chemical products are used to flavor our food, thin paint, remove glue, glue things together, and make us smell good. We depend on these materials to make our lives comfortable.

It is also common to feel safe and secure in our homes. A prevalent idea is that the products used in our homes are safe. Exactly how safe are these products? They are safe if used correctly, and safe if used by knowledgeable people who know which materials catch fire easily and which materials will burn. Many of the solvents, such as alcohol and acetone, are materials that will burn easily, and many aerosol sprays are highly flammable.

Acids and bases are compounds that are caustic to skin. Many products, especially cleaning products, are either acidic or basic. Basic cleaning products utilize the reaction of a base with fat to produce a soap. (Fats are often on surfaces that need to be cleaned, especially in the kitchen.) Thus not only is the grease, a fat, removed but some of it is changed into soap to help the cleaning process. Most commercial detergents contain ingredients to tie up metallic ions responsible for hard water. These ingredients are called *builders* and are often very basic, that is they have a very high pH. Some examples of builders are sodium tripolyphosphate ($Na_5P_3O_{10}$), sodium pyrophosphate ($Na_4P_2O_7$), sodium carbonate (Na_2CO_3), and sodium silicate (Na_2SiO_3).

Drain cleaners for the kitchen are also basic. Bases react with the grease in kitchen clogs to form soap, which helps free the clog. Drain cleaners intended for use in the bathroom contain acids to break down proteins in hair.

The concentration of hydrogen ions can be measured in molarity (M), or moles of H^+ per liter of solution, like most other ions in solution. The H^+ concentration can vary over a wide range, chemists use a convenient scale called pH to denote this range. The pH is defined as the negative log of the molar concentration of the hydrogen ion. Thus, pH is the measure of acidity or basicity. A neutral solution has a pH of 7. pH values less than 7 indicate an acidic solution, with each unit change denoting a tenfold change in acidity. pH values greater than 7 indicate a basic solution.

pH	H^+ concentration (M)	Interpretation
1	10^{-1}	Very acidic
4	10^{-4}	Acidic
7	10^{-7}	Neutral
10	10^{-10}	Basic
13	10^{-13}	Very basic

Consumers should be aware of the dangers involved in using commercial products. Part of this knowledge comes from reading the label and being aware of which ingredients are potentially harmful. The federal government keeps statistics on various types of accidents. There are 25,000 accidents in the home annually in the United States. In these home accidents, 1,400 children are killed and 14,000 children are burned. Consumer products are involved in accidents both inside and outside the home that kill 21,000 people annually and injure 29,000,000.

WASTE AND THE ENVIRONMENT

Paint thinner will contaminate water. It should be completely burned. The other solutions are not toxic.

⚠ *CAUTION*s warn about safety hazards.
*EXTRA*s give helpful hints, additional information, or interesting facts.

Reagents
    ethanol [CH_3CH_2OH]
    acetone [CH_3COCH_3]
    sodium tripolyphosphate [$Na_5P_3O_{10}$]
    sodium pyrophosphate [$Na_4P_2O_7$]
    sodium carbonate [Na_2CO_3]
    sodium silicate [Na_2SiO_3]
    pH test paper

Common Materials
vanilla extract
nail polish remover
cologne
aftershave lotion
paint thinner
oven cleaner
hair spray
toilet bowl cleaner
hand dishwashing detergent
machine dishwashing detergent
bleach
vinegar
drain cleaner
ammonia cleaning product
other products from home
matches

Laboratory Equipment
    evaporating dish
    50 mL beaker
    stirring rod
    laboratory burner

PROCEDURE

1. For each sample listed below, pour no more than 3 mL onto an evaporating dish, one at a time.

    Ethanol
    Acetone
    Vanilla extract
    Nail polish remover
    Cologne
    Aftershave lotion
    Paint thinner
    Oven cleaner
    Hair spray (Spray the evaporating dish heavily.)

⚠ *CAUTION*
Don't burn yourself.
Don't spill these
liquids on yourself or
others. Don't spray
toward an open
flame.

2. First try to ignite the sample with a match. If that doesn't work, tip the Bunsen burner so that its flame touches the sample to try to ignite it.

3. Dissolve 1 g of each of the following materials in 50 mL of water. Record the pH of each solution by touching it with a clean glass stirring rod and then touching a piece of pH paper. The color chart on the container will indicate the pH.

    Toilet bowl cleaner
    Hand dishwashing detergent
    Machine dishwashing detergent
    Bleach
    Sodium tripolyphosphate ($Na_5P_3O_{10}$)
    Sodium pyrophosphate ($Na_4P_2O_7$)
    Sodium carbonate (Na_2CO_3)
    Sodium silicate (Na_2SiO_3)

⚠ *CAUTION*
Don't spill these
materials on yourself
or others. Don't mix
the solutions.

4. For the following liquid materials, test and record the pH.
    Ammonia cleaning product
    Vinegar
    Oven cleaner
    Drain cleaner
    Other products from home

pH AND FLAMMABILITY
PRE-LAB QUESTIONS

Name: _____

Lab Partner: _____

Section: _____   Date: _____

1) In many homes, multiple flammable compounds normally reside under the kitchen sink. Another common storage place for household cleaning supplies and solvents is the "utility" closet that may also house a gas-fueled water heater or furnace. Comment on the risks and wisdom of this practice. Can you suggest alternatives?

2) Assume you are the corporate attorney for a company that produces, packages, and markets multiple household cleaners and solvents. What suggestions would you make to the company CEO concerning consumer warnings?

3) It has been suggested that a labeling system for easy identification of pH and flammability ratings for common substances should be instigated. If this system used color-coding for ease of identifying the characteristics of a substance, what suggestions would you make for the design?

4) Which of the substances we will use in this lab can you find in your own home, apartment, or dorm room?

5) How can you reduce the risk of accidents and injuries involving household chemicals in your domicile?

pH AND FLAMMABILITY
REPORT SHEET

Name: _____

Lab Partner: _____

Section: _____   Date: _____

I. FLAMMABILITY

	Lights with Match	Lights with Stronger Flame
Ethanol	_____	_____
Acetone	_____	_____
Vanilla extract	_____	_____
Nail polish remover	_____	_____
Cologne	_____	_____
Aftershave lotion	_____	_____
Paint thinner	_____	_____
Oven cleaner	_____	_____
Hair spray	_____	_____
Other _____	_____	_____
Other _____	_____	_____
Other _____	_____	_____
Other _____	_____	_____

II. pH

	Brand	pH
Toilet bowl cleaner	_____	_____
Hand dishwashing detergent	_____	_____
Machine dishwashing detergent	_____	_____
Bleach	_____	_____
$Na_5P_3O_{10}$	_____	_____
$Na_4P_2O_7$	_____	_____
Na_2CO_3	_____	_____
Na_2SiO_3	_____	_____
Ammonia cleaner	_____	_____
Vinegar	_____	_____
Oven cleaner	_____	_____

Drain cleaner   _____   _____

Other _____   _____   _____

Other _____   _____   _____

III. QUESTIONS

1. Which products include adequate warnings about flammability?

2. Which products need to include more warnings about flammability?

3. Which of the products tested are unsafe for skin contact? (Products with pH values between 5 and 10 are probably safe for at least short periods of time.)

4. Which of the products tested have adequate warnings concerning pH on the label?

5. Why wouldn't you use machine dishwashing detergent to wash dishes by hand?

6. Why is it suggested to use gloves when using oven-cleaning products?

7. Because of environmental concerns, phosphates have been replaced with other builders, such as sodium carbonate and sodium silicate. Based on the pH of these compounds, does this change make a difference in safety to the consumer?

20 Buffers

Why pH Doesn't Change

OBJECTIVES

To observe the difference between buffer solutions and water in resisting pH change.

To appreciate the importance of buffer systems in maintaining life.

To learn that a weak acid or weak base resists pH change in only one direction.

To demonstrate that a buffer can be overcome by a sufficient quantity of acid or base.

Relates to Chapter 7 of Chemistry for Changing Times, *twelfth ed.*

BACKGROUND

Proper pH is important to many processes and systems. From vegetable gardens to flower colors, pH can alter the quality, color, and other properties of plants. Swimming pools are subject to many foreign substances that can enter the water and alter the pH levels. Rain and the presence of swimmers affect pH. Maintenance of proper pH in pools is critical to the control of algae and bacterial growth as well as comfort for the swimmers. Many enzymes that regulate biochemical processes in the human body will only operate in a very narrow pH range. The optimal pH of human blood is from 7.35 to 7.45. When the pH falls below this range, a condition called *acidosis* results. If the pH rises above the optimal range, the condition is called *alkalosis*. Death may be the result of a blood pH that falls below 6.8, or rises above 7.8. Clearly, the blood buffer system is critical to life. The complex mechanism in human blood that maintains the pH of the blood involves a system of buffers. Buffers work in these and many other applications to maintain specific pH levels.

A *buffer* solution is a solution that will resist large pH changes upon the addition of acid or base to the system. Buffers will also maintain a pH when they are diluted. There are many types of buffers. Some salts, amino acids, and proteins produce natural buffer solutions.

Buffers resist pH change because there is an acid component to neutralize added base and a base component to neutralize added acid. Amino acids and proteins have both acidic and basic functional groups on the same compound, accounting for their usefulness as buffers.

One of the most common types of buffers is the combination of a weak acid and its conjugate base, or a weak base and its conjugate acid. A *conjugate acid-base pair* differs from one another by one H^+ ion. Examples are ammonia (NH_3) and the ammonium ion (NH_4^+), nitrous acid (HNO_2) and the nitrite ion (NO_2^-), and acetic acid (CH_3COOH) and the acetate ion (CH_3COO^-). The acid-base pair that is one of the major components of the buffering system in the blood is the carbonic acid/bicarbonate ion pair (H_2CO_3/HCO_3^-).

In this investigation, it will be demonstrated that a buffer solution requires more acid or base to change its pH than does water. It will also be shown that a weak acid will resist a pH change when base is added, because the acid neutralizes the added base, but the addition of a strong acid will cause a significant pH change, since there is nothing to neutralize the added acid.

Dilution with water does not cause a significant change in pH because the ratio of acid to conjugate base is not changed.

WASTE AND THE ENVIRONMENT
The solutions in this investigation are not toxic. However, acids and bases need to be neutralized or diluted before they are discarded.

▲ *CAUTION*s warn about safety hazards.
*EXTRA*s give helpful hints, additional information, or interesting facts.

Reagents
      1 M acetic acid [CH₃COOH]   wide-range pH paper

Reagents
      1 M acetic acid [CH_3COOH]   wide-range pH paper
      1 M sodium acetate [CH_3COONa]   1 M hydrochloric acid [HCl]
      1 M sodium hydroxide [$NaOH$]
      1 M acetic acid and 1M sodium acetate (a buffer solution)

Laboratory Equipment
      5 extra 50 mL beakers   10 mL graduated cylinder

PROCEDURE
1. Set up four small beakers. Place 10 mL of water in beaker 1. Place 10 mL of 1 M acetic acid (CH_3COOH) in beaker 2. Place 10 mL of 1 M sodium acetate (CH_3COONa) in beaker 3. Place 10 mL of a solution of 1-M acetic acid and 1 M sodium acetate in beaker 4.

2. Measure and record the pH of each solution using pH paper.

3. Add 1 drop of 1 M hydrochloric acid (HCl) to each beaker. Measure and record the pH of the solution in each beaker.

4. Add a second drop of 1 M HCl to each beaker. Measure and record the pH of the solution in each beaker.

5. Add 18 additional drops (a total of 20) of 1 M HCl to each beaker. Measure and record the pH of the solution in each beaker.

6. Add 1 mL of 1 M HCl to each beaker. Measure and record the pH of the solution in each beaker. Repeat until 11 mL have been added to each beaker.

7. Repeat steps 1 through 6 using 1 M sodium hydroxide (NaOH) in place of hydrochloric acid (HCl).

8. Place 10 mL of the buffer solution in a small beaker. Measure and record the pH. Add 5 mL of water to the beaker. Measure and record the pH. Add 5 mL more water to the beaker. Measure and record the pH.

9. Combine the acidic and basic solutions to neutralize each other. Then flush the solutions down the drain with plenty of water.

EXTRA
Acetic acid is a weak acid. The acetate ion is the conjugate base of acetic acid. The acetic acid/acetate ion solution is a buffer solution.

EXTRA
HCl is a strong acid.

▲ *CAUTION*
HCl is caustic.

EXTRA
There are 20 drops in a milliliter.
EXTRA
NaOH is a strong base.

▲ *CAUTION*
NaOH is caustic.

BUFFERS
PRE-LAB QUESTIONS

Name: _____

Lab Partner: _____

Section: _____   Date: _____

1) When a person exerts a great deal of physical energy, as in running a race or playing one-on-one basketball, one of the by-products of the muscular activity that takes place is lactic acid. If the blood buffering system cannot handle the lowered pH, what condition could result?

2) Many consumer products are buffered to protect the body from dangerous pH changes. Pharmaceutical compounds are usually either basic or acidic and not as often neutral compounds. What systems in the body need the protection of a buffer if the substance is ingested? Injected?

3) Many fertilizers are formulated to help maintain specific pH levels for certain applications. Will making the fertilizer the desired pH be enough to accomplish the result? Explain your reasoning.

4) Baking soda (sodium hydrogen carbonate, $NaHCO_3$) acts as a buffer, as does sodium tetraborate [$Na_2B_4O_7$], also called borax. Both are used as water softeners in laundry applications. Each compound raises the pH of the system and then buffers it to keep it consistent. Both of these compounds are active ingredients in pool maintenance products. The hypochlorite ion (ClO^-) is the ingredient used to "shock" a pool. What happens to the pH of the pool?

5) The hypochlorite ion is also the active ingredient in chlorine bleach. Why do people spend large sums of money for pool products that are labeled with cute names like "Alk-Up®" or "SoftSwim®" (actually, $Na_2B_4O_7 \bullet 10H_2O$) instead of buying the substances off grocery store shelves?

BUFFERS
REPORT SHEET

Name: _____

Lab Partner: _____

Section: _____   Date: _____

I.

	Water	Acetic Acid	Acetate Ion	Acetic Acid–Acetate Buffer
	pH	pH	pH	pH
1 drop HCl	_____	_____	_____	_____
2 drops HCl	_____	_____	_____	_____
20 drops HCl	_____	_____	_____	_____
2 mL HCl	_____	_____	_____	_____
3 mL HCl	_____	_____	_____	_____
4 mL HCl	_____	_____	_____	_____
5 mL HCl	_____	_____	_____	_____
6 mL HCl	_____	_____	_____	_____
7 mL HCl	_____	_____	_____	_____
8 mL HCl	_____	_____	_____	_____
9 mL HCl	_____	_____	_____	_____
10 mL HCl	_____	_____	_____	_____
11 mL HCl	_____	_____	_____	_____

II.

	Water	Acetic Acid	Acetate Ion	Acetic Acid–Acetate Buffer
1 drop NaOH	_____	_____	_____	_____
2 drops NaOH	_____	_____	_____	_____
20 drops NaOH	_____	_____	_____	_____
2 mL NaOH	_____	_____	_____	_____

3 mL NaOH   _____   _____   _____   _____

4 mL NaOH   _____   _____   _____   _____

5 mL NaOH   _____   _____   _____   _____

6 mL NaOH   _____   _____   _____   _____

7 mL NaOH   _____   _____   _____   _____

8 mL NaOH   _____   _____   _____   _____

9 mL NaOH   _____   _____   _____   _____

10 mL NaOH   _____   _____   _____   _____

11 mL NaOH   _____   _____   _____   _____

III. Buffer pH   _____   5 mL water added _____   10 mL water added _____

IV. QUESTIONS

1. Which is the best at resisting pH change when acid is added: water, acetic acid, acetate ion, or buffer?

2. Which is the worst at resisting pH change when acid is added: water, acetic acid, acetate ion, or buffer?

3. Which is the best at resisting pH change when base is added: water, acetic acid, acetate ion, or buffer?

4. Which is the worst at resisting pH change when base is added: water, acetic acid, acetate ion, or buffer?

5. Why does the buffer solution experience a large change in pH after 10 mL of acid or base is added?

6. Why would it be useful to "buffer" an aspirin tablet? (Aspirin is acetylsalicylic acid.)

21 *Oxidation and Reduction*

Those Traveling Electrons

OBJECTIVES

To observe several redox reactions and note the changes of substances from their elemental form to their ionic form and vice versa.

To understand that electrochemical cells can either produce a current, or use a current to accomplish a desired process.

To be introduced to the concept of reduction potential.

To construct a voltaic cell and measure the current it produces.

Relates to Chapter 8 of Chemistry for Changing Times, *twelfth ed.*

BACKGROUND

Oxidation-reduction (redox) reactions occur when electrons are transferred from one element to another. Many everyday reactions are based on redox reactions. For example, rust is produced by a redox reaction:

$$4\,Fe(s) + 3\,O_2(g) \;\rightarrow\; 2\,Fe_2O_3(s) \text{ (rust)}$$

One common misconception is that water causes rust. Although objects in water will rust more quickly, water only provides the medium for the electrons to travel more easily.

Redox reactions occur because one element has a stronger attraction for electrons than does another element. The relative attraction for electrons that is exhibited by different elements is called a *reduction potential*. A table listing substances in order of reduction potentials is a useful tool in predicting and designing *electrochemical cells*. An electrochemical cell can be designed to produce a current, as in batteries, or to use an outside current to deposit metals from a solution, as in the process of silver-plating an object of jewelry or dinnerware. In this investigation, we will observe several kinds of electrochemical reactions.

As an example of metal deposition, the Cu^{2+} ion will take electrons from aluminum metal (atoms), producing copper metal (atoms) and aluminum ions:

$$3\,Cu^{2+} + 2\,Al(s) \;\rightarrow\; 2\,Al^{3+} + 3\,Cu(s)$$

Another redox reaction occurs between zinc and iodine in an alcohol solution:

$$Zn(s) + I_2 \,(alcohol) \;\rightarrow\; Zn^{2+} + 2\,I^-$$
$$\text{Purple} \qquad\qquad \text{Colorless}$$

Chlorine will react with the iodide ion in an alcohol solution:

$$2\,I^- + Cl_2 \rightarrow 2\,Cl^- + I_2 \,(alcohol)$$
$$\text{Colorless} \qquad\qquad \text{Purple}$$

Batteries are based on redox reactions that are arranged so that the electrons flow through an external circuit. Different metals can be used to set up a redox cell that will produce a voltage between the two metals. This arrangement is called a *voltaic cell*.

A simple voltaic cell is produced when two metals (coins work well) with different affinities for electrons (reduction potentials) are connected by an electron pathway. Paper towels that have been soaked in a solution of table salt can serve as an electron pathway. The free sodium and chloride ions in the wet towel allow electrons to travel from one metal to another if there is also a connection to carry the electrons back. This is because the electrons cannot build up on one coin. In other words, there must be a complete circuit. A voltage meter can serve as the other connection. The voltage reading is a measure of the difference in the reduction potentials of the metals.

Any ionic solution (one containing dissolved charged particles) will work to support the current between two metals that have differing reduction potentials. Some novelty shops sell clocks that can be run from the acidic environment of any fruit. For this reason, any metal container that may hold liquids with ions in them is subject to becoming part of a voltaic cell and subsequent corrosion.

The life expectance of an iron pipe can be extended by a process called *galvanization*. A galvanized pipe has a zinc coating on it so that the zinc, rather than the iron, provides the electrons to another substance in the environment that might accept them. In this way, the iron is protected from corrosion, at least until all the zinc is gone.

WASTE AND THE ENVIRONMENT

The solutions in this investigation are not toxic. Wet zinc dust in air can burst into flames. Placing the zinc solutions or damp zinc solids on a metal pan protects from fire damage and sets up conditions for an oxidation-reduction reaction that will form zinc oxide, which has a low toxic-hazard rating.

▲ *CAUTION*s warn about safety hazards.
*EXTRA*s give helpful hints, additional information, or interesting facts.

Reagents
    mossy zinc
    tincture of iodine
    1 M copper(II) sulfate [$CuSO_4$]
    1 M sodium chloride [NaCl]
    1 M acetic acid [CH_3COOH]

Common Materials
    aluminum foil
    coins, different metals
    paper towels
    bleach

Laboratory Equipment
    3 50 mL beakers
    voltmeter

PROCEDURE

1. Place several pieces of zinc metal into a small beaker. Cover the metal with tincture of iodine solution. Note the color of the solution and record it. Set it aside.

*EXTRA
A tincture is a
solution in which
alcohol is the solvent.*

2. Place 25 mL of 1 M copper(II) sulfate ($CuSO_4$) in a 50 mL beaker. Roll a 4-in. by 4-in. piece of aluminum foil into a roll. Place the roll in the beaker. It does not have to be completely immersed. Set it aside.

3. Place 5 mL of 1-M copper(II) sulfate into a second 50 mL beaker and set it aside.

4. Obtain two coins made of different metals. Cut several pieces of paper towel to the size of the coins. Wet these pieces with 1-M sodium chloride (NaCl). Place the pieces of wet paper towel on top of one coin and then place the other coin on top of the stack. Ask the instructor to measure the voltage produced by this voltaic cell and record the value.

5. If the purple color of the solution (prepared in step 1) over the zinc has faded, pour the liquid into a second beaker, leaving the metal behind. Add several drops of bleach to the solution. Bleach contains Cl_2 molecules. If a solid, $[Zn(OH)_2]$, appears, add a few drops of 1-M acetic acid (CH_3COOH). Record any changes you observe.

⚠ *CAUTION*
*Bleach and acetic
acid are caustic.*

6) Remove the aluminum foil from the beaker of copper sulfate (step 2). Record any changes you observe. The black solid is copper metal. Observe the difference in the aluminum foil. Compare the colors of the solutions in the two beakers of copper sulfate. Record your observations.

7. Place another piece of aluminum foil in the other copper solution (step 3) to remove the copper, then flush the solutions down the drain with plenty of water. Throw away the aluminum foil in the trash. Wash and save the coins (step 4). Place the zinc solutions from steps 1 and 5 on a metal pan to dry. The zinc oxide formed can then be buried in a landfill.

OXIDATION AND REDUCTION
PRE-LAB QUESTIONS

Name: _____

Lab Partner: _____

Section: _____   Date: _____

1) Have you ever experienced a little jolt while a dental assistant was cleaning your teeth? It happens when the metal instrument he or she is holding touches a metallic filling, or more often, the metal base to a crown. It is called *galvanic shock*. Explain why this happens.

2) Some people will test a small 9-volt battery to see if it is still usable by touching the terminal end of it to their tongue. (Not a wise practice, by the way.) Why doesn't it work just as well to lay a finger across the two terminals?

3) Explain why care must be taken when serving food from sterling silver if the food is acidic.

4) Consider a galvanized iron pipe full of dirty water that may contain ions of other metals. These metal ions might have reduction potentials compared to that of the iron so that a voltaic cell is produced that would corrode the pipe. The presence of zinc may prevent the corrosion. On a table of reduction potentials, would you expect zinc to fall between iron and the other metal, or would zinc be located either above or below both iron and the other metal? Explain your reasoning.

OXIDATION AND REDUCTION
REPORT SHEET

Name: _____

Lab Partner: _____

Section: _____   Date: _____

I. VOLTAIC CELL VOLTAGE   _____ V
    INSTRUCTOR'S INITIALS   _____

    1. Would other coins work?

    2. Would pieces of metal work?

    3. Would other ionic compounds like KBr work instead of NaCl?

II. ZINC, IODINE, CHLORINE

	Zinc/Iodine	Copper(II) Sulfate/Aluminum	Copper(II) Sulfate
Initial color	_____	_____	_____
Final color	_____	_____	_____
Color after addition of zinc/iodine	_____		
Color after addition of bleach	_____		
Color after addition of 1-M CH_3COOH (if necessary)	_____		

    1. Compare the aluminum foil pieces.

    2. Compare the beakers of copper(II) sulfate.

    3. Is there zinc in the first solution? In what form?

171

4. In what form is the zinc after the purple color has disappeared from the solution?

5. Does bleach change the zinc back to the metal?

6. In what form is the chlorine after the purple color has disappeared from the solution?

III. COPPER AND ALUMINUM

1. What does the difference in color of the copper solutions indicate?

2. What happened to the aluminum foil?

3. In what form is the copper at the end of the investigation?

4. Could copper be recycled by collecting it on aluminum foil?

22 *Single Replacement Reactions*

Chickens Have a Pecking Order Too

OBJECTIVES

To illustrate the concept of an activity series and demonstrate its usefulness in predicting exchange reactions.

To produce an activity series based on reactions between six common metals and aqueous solutions of their nitrate salts.

To observe the changes that indicate a chemical reaction has occurred.

To practice the representation of exchange reactions as net ionic equations.

Relates to Chapter 8 and 5 of Chemistry for Changing Times, *twelfth ed.*

BACKGROUND

Reactions in which one element takes the place of another in a compound are referred to as *single replacement* reactions. These reactions occur according to a type of hierarchy among the elements. This order is determined by the general reactivity of the elements, that is, by how easily they gain or lose electrons to form compounds. Some elements are more stable when they are in their elemental form, while others are more stable when they are bonded to other atoms, or are in their ionic form. Some elements have a greater attraction for their electrons than do ions of other elements. Other ions have enough attraction for electrons to remove them from another element. The relative attraction of the elements for electrons determines which ones are more likely to be incorporated into compounds, reacting with water, steam, acids, or other ions, and which are more likely to exist in their pure, elemental form.

Since one substance begins in its elemental form and ends up as part of a compound, all single replacement reactions can also be classified as *oxidation-reduction* reactions. This is because *oxidation states* of the elements are changed. For example, aluminum metal is more stable when bonded to the sulfate ion than is copper. Therefore, the aluminum atoms on an aluminum wire dropped into a copper sulfate solution will be *oxidized* as the wire disintegrates. At the same time the copper metal is *reduced* and is deposited onto the remaining wire. The single replacement reaction is represented as follows.

$$3CuSO_4\ (aq)\ + 2Al\ (s)\ \rightarrow Al_2(SO_4)_3\ (aq)\ + 3Cu\ (s)$$

Because the sulfate ion (SO_4^{2-}) does not actually take part in the reaction, it can be left out of the completed chemical equation, leaving a *net ionic* equation as follows.

$$3Cu^{2+}\ (aq)\ + 2Al\ (s)\ \rightarrow 2Al^{3+}\ (aq)\ + 3Cu\ (s)$$

A net ionic equation only shows the substances that are involved in the reaction.

A ranking of the elements according to their tendency to form compounds is called an *activity series*. There is an activity series for metals and another for nonmetals. The nonmetal series usually incorporates the halogens. Because of periodic trends in reactivity, this nonmetal series is in the same order as the elements appear proceeding down the group on the periodic table.

The metallic activity series includes the elements from groups 1 and 2, as well as the transition metals and the metals that appear in groups 11 through 13. This series is used to predict the viability and relative speed of replacement reactions between metals, or between metals and metal ions. The series was developed based upon experimental evidence from reactions which occur between the metals and other substances both in nature and in the lab, the speed of these reactions, and the frequency with which they occur. The approximate order of the metals in this series is predictable based upon knowledge of periodic trends in reactivity and ionization energy. The group 1 and 2 metals tend to cluster at the top of the series, being the most reactive, while the precious metals that are sometimes found as free elements in nature fall at the end of the series.

A reaction is predicted with the knowledge that between two metals, the one that is higher in the activity series will tend to exist in the bonded, or ionized, state while the one that is lower in the series will tend to exist in its un-bonded, or un-ionized, elemental form. As would be expected from the reaction in the example above, aluminum sits higher in the activity series than does copper, as shown in this abbreviated activity series.

Lithium
Aluminum
Cadmium
Copper
Gold

The speed of a reaction can also be predicted by the fact that the further apart two elements are in the series, the faster a reaction between them will begin to be evident.

In this investigation, we will use six common metals and their respective aqueous nitrate solutions. Using a series of reactions, we will develop a point system for each element's ability to replace other elements in the nitrate compounds and the speed with which these replacements begin to take place. An abbreviated activity series can be derived from this point system.

WASTE AND THE ENVIRONMENT

It is usually not good to put metal ions into the environment. In this experiment, the amount of material will be so small that they will not have a measurable effect on the environment.

⚠ *CAUTION*s warn about safety hazards.
*EXTRA*s give helpful hints, additional information, or interesting facts.

Reagents
  1 M calcium nitrate [$Ca(NO_3)_2$]
  1 M copper(II) nitrate [$Cu(NO_3)_2$]
  1 M iron(II) nitrate [$Fe(NO_3)_2$]
  1 M magnesium nitrate [$Mg(NO_3)_2$]
  1 M tin nitrate [$Sn(NO_3)_2$]
  1 M zinc nitrate [$Zn(NO_3)_2$]
  1 M nitric acid [HNO_3]
  calcium metal pieces, small
  copper metal wire, small
  iron metal pieces, small
  magnesium metal ribbon, small
  tin metal pieces, small
  zinc metal pieces, small

Laboratory Equipment
  7 small test tubes, one for each metal tested

PROCEDURE

1. Label each of the seven test tubes with one of the following labels: Zn^{2+}, Fe^{2+}, Ca^{2+}, Mg^{2+}, Sn^{2+}, Cu^{2+}, and HNO_3.

2. Place approximately 1 mL of the nitrate solution of the indicated ion in each of the test tubes.

3. Your instructor will assign a metal (or perhaps two metals) to you for testing in each solution. Place a small amount of the metal assigned to you in each of the marked test tubes containing the solutions. Check the time on a clock and begin watching them to see when reactions are evident. Check them at 5 minutes, at 30 minutes, and then again at the end at the lab period. (Be sure to start all reactions at the same time.)

4. Record your results in the table provided on the report sheet. (Be sure to note the changes you observe that indicate a reaction is taking place.)

5. For each test tube in which you detected a reaction, write a complete and balanced reaction equation on the report sheet.

6. Get the results from the other lab groups and fill in the complete activity series determination chart on the report page. Assign 3 points to a metal that reacted in under five minutes; 2 points to a reaction that occurred within 30 minutes; 1 point to one that reacted within 2 hours; and a "0" to one that was not evident at the end of the lab period. Add the total points (vertically) for each element.

7. Dispose of the contents of each test tube by flushing them down the drain with plenty of water, unless otherwise directed by your instructor.

⚠ CAUTION
Nitric acid is corrosive to living tissue.

EXTRA
HNO_3 is a source of $H^=$.

EXTRA
Remember that the indications of a chemical reaction are color changes, bubbles, temperature changes, and the formation of precipitates.

SINGLE REPLACEMENT REACTIONS
PRE-LAB QUESTIONS

Name: _____

Lab Partner: _____

Section: _____   Date: _____

1) Platinum is a precious metal used in the finest jewelry. Where on the activity series would you expect to find it? Explain your reasoning.

2) Sodium metal is near the top of the activity series and therefore reacts easily with moisture in the air, oxygen, and weak acids. Devise a storage method for elemental sodium.

3) Use the activity series on the report page to predict which of the following combinations will result in a reaction. Write "yes" or "no" in the blank before each combination.

_____ a) metallic iron in a solution of lithium nitrate

_____ b) metallic zinc in a solution of copper sulfate

_____ c) metallic strontium in a solution of nickel chloride

4) Which of the following two combinations would you expect to produce the fastest reaction?

_____ a) chromium and cadmium nitrate

_____ b) barium and lead nitrate

SINGLE REPLACEMENT REACTIONS
REPORT SHEET

Name: _____

Lab Partner: _____

Section: _____   Date: _____

I. REACTION TIMES AND CHANGES OBSERVED
First METAL Assigned: _____

Time	HNO_3	$Zn(NO_3)_2$	$Fe(NO_3)_2$	$Ca(NO_3)_2$	$Mg(NO_3)_2$	$Sn(NO_3)_2$	$Cu(NO_3)_2$
within 5 minutes							
30 minutes							
2^+ hours							
never							

Second METAL Assigned: _____

Time	HNO_3	$Zn(NO_3)_2$	$Fe(NO_3)_2$	$Ca(NO_3)_2$	$Mg(NO_3)_2$	$Sn(NO_3)_2$	$Cu(NO_3)_2$
within 5 minutes							
30 minutes							
2^+ hours							
never							

II. ACTIVITY SERIES DETERMINATION

	Zn/Zn^{2+}	Fe/Zn^{2+}	Ca/Zn^{2+}	Mg/Zn^{2+}	Sn/Zn^{2+}	Cu/Zn^{2+}
Time						
Points						
	Zn/Fe^{2+}	Fe/Fe^{2+}	Ca/Fe^{2+}	Mg/Fe^{2+}	Sn/Fe^{2+}	Cu/Fe^{2+}
Time						
Points						
	Zn/Ca^{2+}	Fe/Ca^{2+}	Ca/Ca^{2+}	Mg/Ca^{2+}	Sn/Ca^{2+}	Cu/Ca^{2+}
Time						
Points						
	Zn/Mg^{2+}	Fe/Mg^{2+}	Ca/Mg^{2+}	Mg/Mg^{2+}	Sn/Mg^{2+}	Cu/Mg^{2+}
Time						
Points						
	Zn/Sn^{2+}	Fe/Sn^{2+}	Ca/Sn^{2+}	Mg/Sn^{2+}	Sn/Sn^{2+}	Cu/Sn^{2+}
Time						
Points						
	Zn/Cu^{2+}	Fe/Cu^{2+}	Ca/Cu^{2+}	Mg/Cu^{2+}	Sn/Cu^{2+}	Cu/Cu^{2+}
Time						
Points						
	Zn/H^{+}	Fe/H^{+}	Ca/H^{+}	Mg/H^{+}	Sn/H^{+}	Cu/H^{+}
Time						
Points						
	Zn Total	Fe Total	Ca Total	Mg Total	Sn Total	Cu Total
Total Points						

III. QUESTIONS

1. How useful would you say the activity series is for aiding chemists in knowing the probable outcome of a reaction involving two metals?

2. Write a balanced equation below for each of the reactions for the first metal you were assigned. (Give the reactants for each attempted reaction. For those in which nothing was visible, write "no reaction" following the yield sign.)

3. List the metals we tested in order of activity, with the most active metal first and the least active metal last. (For example, the metal with the most points would be the most active and be placed at the beginning of the list.)

4. How well does this agree with the accepted order of the activity series: calcium, magnesium, zinc, iron, tin, copper?

5. What factors could explain any discrepancies you find?

6. According to the data for our class, where in the series would you place hydrogen?

7. Magnesium and calcium are not expected to react with themselves, and yet there was a definite reaction. What was happening?

23 *Silver Tarnish Removal*

Save the Elbow Grease

OBJECTIVES

To understand the process of electron transfer in a reaction as an ion changes to its elemental form.

To understand the process of tarnishing of silver objects.

To produce a reaction in which a replacement reaction occurring as an oxidation-reduction reaction is useful as a household application of chemistry.

To observe the removal or sulfur ions from tarnished silver and the production of aluminum sulfide as a by-product.

Relates to Chapter 8 of Chemistry for Changing Times, *twelfth ed.*

BACKGROUND

Silver is a metal that is often used as currency and to make beautiful ornaments, jewelry, flatware, and hollowware. (Flatware and hollowware are types of dining utensils.) Although silver does not react with oxygen in the air, it does react with sulfur to produce silver sulfide (Ag_2S), which is black and better known as tarnish.

Sulfur is present in the air and in foods. Bad-smelling sulfur compounds called mercaptans are added to natural gas to alert people to a gas leak. Natural gas does not have much odor in the absence of the sulfur compounds. The mercaptans are necessary for safety but cause silver to tarnish more rapidly in homes using natural gas.

Silver also tarnishes if it is wrapped with a rubber band. Most rubber materials have been vulcanized and thus contain sulfur. In the process of vulcanization, a mixture of rubber and sulfur is heated. The sulfur cross-links the long polymer molecules, giving the rubber stability and strength.

Other substances will also cause silver to tarnish. Eggs are an example of a food that contains sulfur and will cause tarnish on silver. The rotten-egg smell is hydrogen sulfide gas (H_2S). When dissolved in water, the compound thioacetamide also produces hydrogen sulfide.

One way to remove the silver tarnish is by abrasion. The tarnish is physically removed by rubbing with a rough or abrasive material. The silver that is part of the silver sulfide tarnish is removed from the silver object, so each time an object is polished a small amount of silver is lost.

An electrochemical reduction (oxidation-reduction reaction) is another way to get rid of the tarnish. In this method, the silver in the tarnish is changed back to silver metal so no silver is lost. A necessary requirement for the electrochemical reduction is another metal, such as aluminum or zinc, that will give up electrons to silver. Copper can be cleaned in the same manner. The reaction for silver is

$$3\ Ag_2S(s)\ +\ 2\ Al(s)\ \rightarrow\ Al_2S_3(s)\ +\ 6\ Ag(s).$$

S Affinity for Al > Ag

In order to accomplish the reaction, salt (sodium chloride) and baking soda (sodium bicarbonate) are added to hot water. The salt provides electrolytes (ions that will carry the charge or electrons). Sodium bicarbonate removes the aluminum oxide coating from the aluminum and supplies additional electrolytes. The aluminum foil serves as the source of the electrons given up to the silver ions of the silver sulfide. This reaction produces silver metal. The negative sulfur ions are then attracted to the newly formed aluminum ions. Thus, the silver loses its "tarnish" to the aluminum. Best results are obtained with good-quality silver, such as flatware or jewelry.

This same procedure can be done at home with a Pyrex or Corningware dish (not a metal one), salt, baking soda, and aluminum foil. Large pieces can be cleaned, half at a time, as long as the dish is large enough to accommodate them. Cleaning your silver at home using household aluminum foil is cheaper than buying one of the available kits.

WASTE AND THE ENVIRONMENT

The tarnish-removal solution is not toxic. The thioacetamide will cause tarnish and should not be put into the water system.

⚠ *CAUTION*s warn about safety hazards.
*EXTRA*s give helpful hints, additional information, or interesting facts.

Reagents
     sodium bicarbonate [$NaHCO_3$]
     thioacetamide [CH_3CSNH_2]
     sodium chloride [$NaCl$]

Common Materials
     tarnished silver
     aluminum foil
     bleach (sodium hypochlorite)
     paper towels
     liquid soap

Special Equipment
     2000 mL beaker
     ring stand, ring, and wire gauze
     forceps or tongs
     laboratory burner

PROCEDURE

1. Bring to lab a tarnished piece of silver. This may be jewelry, flatware, hollowware, or any solid silver or silver plate object. It should not have gems attached. If you cannot supply a piece of tarnished silver, the lab instructor will have silver for you to tarnish, clean, and return.

2. If your silver is not tarnished, it can be tarnished by immersing it in a thioacetamide solution. To prepare the solution, put 6.5 g of thioacetamide (CH_3CSNH_2) in 100 mL of water. Heat the solution to about 70°C. Make sure that all of the crystals dissolve because they will pit silver. Place the silver object in the hot solution. Within a few minutes, tarnish should appear on the silver object. Remove the silver object from the solution. Rinse the silver object with water and set it aside to dry.

CAUTION
Don't burn yourself. The hot solution contains sulfide ions. Because of the smell and toxicity of hydrogen sulfide, heat the solution in the hood. One solution will suffice for the entire lab.

3. Line the bottom and side of a 2-L beaker with aluminum foil.

4. Place about 1.7 L of water in the 2-L beaker. Add 13 g of sodium bicarbonate ($NaHCO_3$) and 13 g of sodium chloride ($NaCl$).

EXTRA
$NaHCO_3$ is also called baking soda. $NaCl$ is table salt.

5. Place the tarnished silver in contact with the aluminum foil and completely submerge it.

6. Heat the solution to a simmer (just before a boil) with a ~~Bunsen burner.~~ *Hot Pl.* Continue to heat for ten minutes. Notice the sulfur smell.

CAUTION
Don't burn yourself.

EXTRA
Doesn't that look nice?

7. Remove the Bunsen burner. Remove the silver object from the beaker. Rinse it and dry it with a paper towel.

8. If the silver looks dull or yellowish, rub it lightly with a small amount of sodium bicarbonate or liquid soap on your damp fingers.

EXTRA
The solution will become very warm if the addition is too rapid.

9. Flush the tarnish-removal solution down the drain with plenty of water. The thioacetamide solution should be added slowly to a solution of sodium hypochlorite (bleach). When the solution has cooled to room temperature, it should be flushed down the drain with plenty of water unless otherwise directed by your instructor.

SILVER TARNISH REMOVAL
PRE-LAB QUESTIONS

Name: _____

Lab Partner: _____

Section: _____   Date: _____

1) List foods that probably should not be placed directly on silver serving dishes.

   Eggs
   Onions
   Garlic

2) Silver stored in some plastics tends to tarnish quickly. What would you suspect is the reason?   *S used in plastic production*

3) How could silver be protected from tarnish in homes with gas-fueled furnaces and ranges?   *Cover it*

4) Some lotions and cosmetic products will cause silver to tarnish when they come in contact with it. Why might this happen?

   Contain S

SILVER TARNISH REMOVAL
REPORT SHEET

Name: _____

Lab Partner: _____

Section: _____   Date: _____

I. INSTRUCTOR'S INITIALS FOR CLEAN SILVER   _____

II. QUESTIONS
1. Look at the aluminum foil. What happened to the sulfur?

2. Would this method be preferable to a paste polish for removing the tarnish from the crevices of a very intricate piece of silver? (Removing the tarnish from the crevices of a very old piece devalues the piece in the eyes of antique collectors.)

3. Why is it necessary for the silver and the aluminum to touch?

4. Is it necessary to be in a chemical laboratory to clean silver?

5. Silver cleaning kits are sold on television and in stores that contain a large plate. The silver to be cleaned has to make contact with the plate. What material do you think is used to make the plate?

6. Why should this method not be used for silver with stones in it?

7. Why would cleaning silver pieces by abrasion eventually cause the loss of detail in the pattern on the piece?

24 *Organic Qual Scheme*

Follow that Roadmap

OBJECTIVES

To learn to perform tests for distinguishing different organic functional groups.

To demonstrate the difference in solubility of functional groups in water.

To identify the functional group of an unknown sample.

To understand the usefulness of a qualitative analysis (qual) scheme in analytical chemistry.

Relates to Chapter 9 of Chemistry for Changing Times, *twelfth ed.*

BACKGROUND

The functional group of an organic molecule is a specific arrangement of atoms within a molecule and is the site where reactions occur for the compound. These functional groups give each class of organic compounds their characteristic properties. Furthermore, the different functional groups can be classified by their reactions with various chemicals. For instance, all alcohols containing ten carbon atoms or fewer will react with cerric ammonium nitrate to produce a color change from yellow to red.

Another way to distinguish the different functional groups is to observe their solubility in water. Compounds containing only carbon and hydrogen (hydrocarbons) such as alkanes and alkenes, are much less soluble than compounds containing oxygen or nitrogen (amines, alcohols, ketones, aldehydes, and carboxylic acids). Many compounds containing nitrogen or oxygen will form hydrogen bonds with water and thus be more soluble than hydrocarbons. *Hydrogen bonds* are strong attractions between polar molecules containing hydrogen bonded to either oxygen or nitrogen. Compounds which are small molecules containing fewer than ten carbons along with oxygen or nitrogen are usually soluble in water. Large hydrogen-bonding molecules are not very soluble, because the *hydrophobic*, or "water fearing," hydrocarbon groups overwhelm the hydrogen bond attraction. Compounds such as alkyl halides are polar and are more soluble than alkanes but are much less soluble than compounds that exhibit hydrogen bonding.

The individual tests for organic groups are specific reactions that occur only with a particular group or related groups. Some examples are listed here:

Beilstein test—When an alkyl halide is placed in a flame on a copper wire, the compound decomposes to form hydrogen halides, which react with copper to form cupric halides, which burn with a green flame.

Cerric nitrate test—Alcohols containing ten carbons or fewer react with cerric ammonium nitrate to form a red complex. The presence of this complex causes the solution to change from yellow to red or reddish brown.

Baeyer test—Aqueous potassium permanganate will oxidize alkenes to *diols*, alcohols with two alcohol functional groups. The purple permanganate is reduced to brown manganese dioxide. This test is not alkene specific, as some alcohols and aldehydes are also oxidized by permanganate.

Iron chloride test—An aqueous solution of iron (III) chloride will turn purple in the presence of many phenols. The color is due to the formation of an iron phenoxide salt.

An organic qual scheme (short for qualitative schematic analysis) allows one to distinguish between specific classes of compounds using a series of tests that are either positive or negative. As each successive test is performed, the possibilities are narrowed until the class of the compound is identified. The order in which these tests are performed is critical to the accuracy of the determination. As you perform the qual scheme with your unknown, trace your steps on the report sheet. If necessary, use separate colors on the chart to keep your unknown determinations straight. The scheme given below is applicable to small organic molecules.

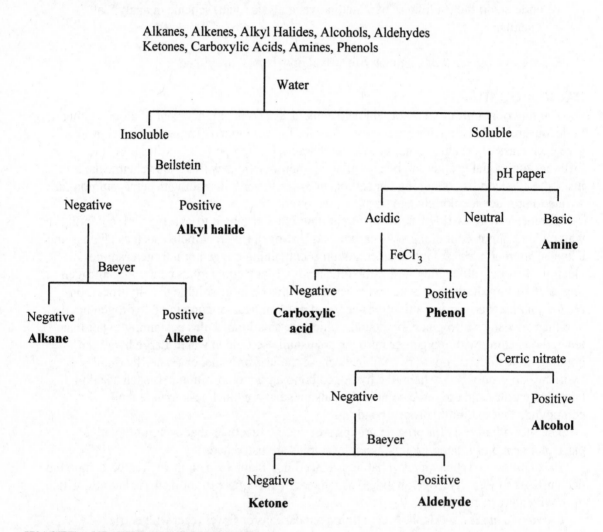

WASTE AND THE ENVIRONMENT

Metal ions should not be allowed to get into the environment. Organic materials can also pollute the environment.

⚠ *CAUTION*s warn about safety hazards.
*EXTRA*s give helpful hints, additional information, or interesting facts.

Reagents

hexane [CH$_3$(CH$_2$)$_4$CH$_3$]	2-chloro,2-methyl propane[CH$_3$C(Cl)(CH$_3$)CH$_3$]
n-octene [CH$_2$=CH(CH$_2$)$_5$CH$_3$]	pH paper
benzaldehyde [C$_6$H$_5$C(O)H]	cerric nitrate solution (4 g in 10 mL 2-M HNO$_3$)
resorcinol [C$_6$H$_4$(OH)$_2$]	1% iron(III) chloride solution [FeCl$_3$]
acetic acid (conc.) [CH$_3$COOH]	copper wire (1 cm lengths)
ethanol [CH$_3$CH$_2$OH]	acetone (as a test reagent) [CH$_3$COOH]
acetone [CH$_3$C(O)CH$_3$]	1% potassium permanganate [KMnO$_4$]
isobutylamine [CH$_3$CH(CH$_3$)CH$_2$NH$_2$]	

Common Materials

food coloring   paper towels

Laboratory Equipment

watch glass   small test tubes
stirring rod

PROCEDURE

1. Before attempting analysis of the unknowns, run the Baeyer, FeCl$_3$, cerric nitrate, Beilstein, pH, and water solubility tests on known compounds that will show both a positive and a negative result so that you will recognize them. Then obtain an unknown and carry out the following tests as indicated in the qual scheme.

EXTRA
Solubility varies from very soluble to insoluble.

2. **Water Solubility Test**
Add 3 drops of unknown to 0.5 mL distilled water in a small test tube. Stir vigorously for one minute and observe the results. An interface between the two layers indicates insolubility. If the unknown is soluble, continue with step 3. If the unknown is insoluble in water, go to step 7. It may be easier to determine solubility if the water is colored with food color and the test tube is slanted.

EXTRA
pH test paper should never be dipped into a test solution.

3. **pH Test**
On a watch glass, test the unknown with pH test paper by using a glass stirring rod to place 1 drop of the unknown onto a small piece of pH test paper. If the unknown is basic, the unknown is an amine. Go to the report sheet. If the unknown is acidic, go to step 4. If the solution is neutral, go to step 5.

4. **FeCl$_3$ Test**
Add 1 drop of unknown to 5 drops of 1% iron(III) chloride (FeCl$_3$) on a watch glass. A purple color indicates the presence of a phenol. The purple may appear as a purple cloud in the yellow FeCl$_3$. The lack of purple color indicates a carboxylic

193

acid. Clean the watch glass with a paper towel and discard it in a waste container. Go to the report sheet.

5. **Cerric Nitrate Test**
Place 5 drops of cerric nitrate solution on a watch glass. Add 2 drops of unknown and stir with a glass rod. A positive test is a color change from yellow to red and indicates an alcohol or phenol. Go to the report sheet. A negative test indicates aldehyde or ketone. Go to step 6. Clean the watch glass with a paper towel. Discard the paper towel in a waste container.

6. **Baeyer Test**
Place 4 drops of acetone and one drop of unknown on a watch glass. Add 1 drop of 1% potassium permanganate ($KMnO_4$) and stir. A positive test is a color change from purple to brown. An aldehyde will produce a positive Baeyer test. A ketone will produce a negative Baeyer test. Clean the watch glass with a paper towel. Discard the paper towel in a waste container. Go to the report sheet.

7. **Beilstein Test**
Heat the tip of a copper wire in the flame of a Bunsen burner until no coloration of the flame is produced. Allow the wire to cool. Working over a watch glass, drop the unknown onto the tip of the wire and place the wire in the flame. A green flame is an indication of an alkyl halide. Go to the report sheet. If there is no green flame, go to step 8. The copper wire can be thrown into the trash unless otherwise directed by your instructor.

8. **Baeyer Test**
Place 4 drops of acetone and 1 drop of unknown on a watch glass. Add 1 drop of 1% potassium permanganate ($KMnO_4$) and stir. A positive test is a color change from purple to brown. An alkene will produce a positive Baeyer test. An alkane will produce a negative Baeyer test. Clean the watch glass with a paper towel. Discard the paper towel in a waste container. Go to the report sheet.

9. Pour the waste from step 1 into a waste container marked "organics and water." Absorb the residue of the other tests in a paper towel and place it in a waste container unless otherwise directed by your instructor.

EXTRA
Cerric nitrate solution is prepared by dissolving 4.0 g of cerric ammonium nitrate $[(NH_4)_2Ce(NO_3)_6]$ in 10 mL of 2 M nitric acid $[HNO_3]$.

⚠ *CAUTION*
Nitric acid is caustic.

⚠ *CAUTION*
Do not burn yourself.
EXTRA
Holding the copper wire with a cork into which it is inserted may be an easy method.

ORGANIC QUAL SCHEME
PRE-LAB QUESTIONS

Name: _____

Lab Partner: _____

Section: _____   Date: _____

1) In finding the identity of an unknown organic compound, determining the class to which it belongs is only the first step. What else would need to be known in order to identify the compound specifically?

2) Only tiny amounts of an unknown are required for each test. Why is this advantageous?

3) If you knew that a compound of the alcohol class was toxic to living cells and made a good antiseptic, could you suspect this is true of other alcohols also? Explain.

ORGANIC QUAL SCHEME
REPORT SHEET

Name: _____

Lab Partner: _____

Section: _____   Date: _____

OBSERVATIONS OF TESTS ON KNOWN COMPOUNDS
  Water solubility

  pH

  $FeCl_3$

  Cerric nitrate

  Baeyer

  Beilstein

QUESTIONS

1. Study the qual scheme and give an example of an incorrect determination that could be made by performing the tests out of order.

2. List an application in which a qual scheme would be useful.

3. Many organic compounds have distinct odors. Why doesn't one simply sniff the unknown to make an identification?

Mark over the dotted lines with a dark line according to your test results to find the classification of the unknown. If more than one unknown was analyzed, use a different color for each unknown and provide a color key.

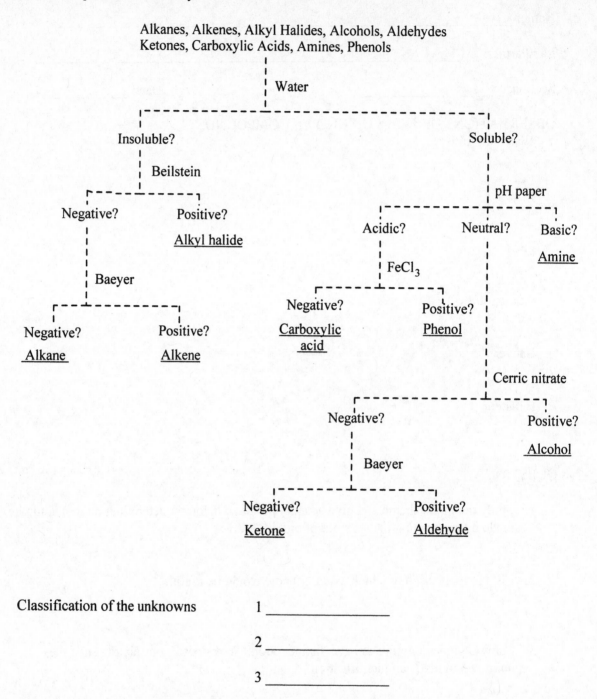

Alkanes, Alkenes, Alkyl Halides, Alcohols, Aldehydes
Ketones, Carboxylic Acids, Amines, Phenols

Water

Insoluble?

Beilstein

Negative?   Positive?
             Alkyl halide

Baeyer

Negative?   Positive?
Alkane   Alkene

Soluble?

pH paper

Acidic?   Neutral?   Basic?
                       Amine

FeCl$_3$

Negative?   Positive?
Carboxylic   Phenol
acid

Cerric nitrate

Negative?   Positive?
                     Alcohol

Baeyer

Negative?   Positive?
Ketone   Aldehyde

.   Classification of the unknowns   1 _____

2 _____

3 _____

INVESTIGATION

25 *Synthesis of Esters*

How Sweet the Smell!

OBJECTIVES

To recognize the functional group characteristic of an ester.

To synthesize several esters and identify the familiar odors of each.

To be able to draw the structures of the reactants involved in an esterification reaction and predict the structure of the resulting ester.

To understand the nature of chemical synthesis and its importance to modern industry and lifestyle.

Relates to Chapters 9 and 17 of Chemistry for Changing Times, *twelfth ed.*

BACKGROUND

Nature contains a large number of compounds that smell sweet to us. Flowers and fruits, in particular, contain compounds whose aroma is pleasing to us.

These compounds are *volatile*, meaning that they vaporize easily, and are therefore able to be detected through our sense of smell. They are also highly potent and only very small amounts are necessary to give the aroma or flavor we recognize from fruits, flowers, and commercial uses in food and other common products. Methyl salicylate is familiar to you as the wintergreen scent and flavor. Ethyl acetate is found in nail polish remover.

One class of compounds that is usually sweet smelling is the *esters*. Chemists have separated, identified, and synthesized many of these compounds, making them available at a reasonable cost for use in cosmetics, household products, cleaners, and even foods. Synthetic molecules are identical to the molecules we find in nature, but the process of isolating "natural" molecules compared to the relative inexpensive process of producing, or synthesizing, large quantities in a lab makes the synthetic molecules much more affordable. These compounds are synthesized by specific chemical reactions. One example is the class of compounds called esters. Esters have the following general formula:

$$\overset{\overset{\displaystyle O}{\displaystyle \|}}{R-C-OR'}$$

where R and R' are alkyl (hydrocarbon) groups. As an example, let R be CH_3, a methyl group, and let R' be $CH_3(CH_3)_7$, an octyl group. The ester produced is called *octyl acetate* and has the aroma of orange peelings.

$$\overset{\overset{\displaystyle O}{\displaystyle \|}}{H_3C-C-O(CH_2)_7CH_3}$$

Octyl Acetate

One method of producing esters is to react a *carboxylic acid* with an *alcohol* in an acidic medium such as sulfuric acid. The alcohol contains a *hydroxyl group* (OH) and the organic, or carboxylic, acid contains a *carboxyl group* (COOH). The carboxyl group can be thought of as a *carbonyl group* (C=O) and a hydroxyl group linked together. The sulfuric acid acts as a catalyst by attaching to the oxygen of the carbonyl group. This makes the carbon of the carboxylic acid group slightly positive so it will attract the oxygen of the alcohol. The hydrogen of the alcohol is then close enough to the OH group of the carboxylic acid to react to form water (H_2O). Now the carbon of the carbonyl group on the acid is bonded to an OR group and an OH group. The R group attracts enough of the electron density so that the H^+ ion is released from the OH group, leaving a carbonyl group. A carbonyl group together with the OR group forms an ester group. The box encloses the atoms that will form the water molecule.

$$\underset{\substack{\text{Carboxylic} \\ \text{Acid}}}{R-\overset{\overset{\displaystyle O}{||}}{C}-\boxed{OH}} \;+\; \underset{\text{Alcohol}}{\boxed{H}OR'} \quad \overset{H^+}{\underset{}{\rightleftharpoons}} \quad \underset{\text{Ester}}{R-\overset{\overset{\displaystyle O}{||}}{C}-OR'} \;+\; \underset{\text{Water}}{H_2O}$$

Although the reaction is reversible, a greater production of ester can be forced by using an excess of alcohol. Combinations of different carboxylic acids and alcohols produce different esters, each of which has a particular aroma. Notice that the carboxylic acids usually do not have a pleasant aroma.

The reaction for producing octyl acetate is as follows.

$$\underset{\text{Acetic Acid}}{H_3C-\overset{\overset{\displaystyle O}{||}}{C}-OH} \;+\; \underset{\text{Octyl Alcohol}}{HO(CH_2)_7CH_3} \quad \overset{H^+}{\underset{}{\rightleftharpoons}} \quad \underset{\text{Octyl Acetate}}{H_3C-\overset{\overset{\displaystyle O}{||}}{C}-O(CH_2)_7CH_3} \;+\; \underset{\text{Water}}{H_2O}$$

The name of the ester is listed in two parts. The group attached to the oxygen is listed first; that group comes from the alcohol. The second name is the acid name with *oic acid* or *ic acid* deleted and *oate* or *ate* added. Thus octyl alcohol and acetic acid form the ester octyl acetate.

WASTE AND THE ENVIRONMENT

Esters are organic substances that will contaminate waterways if flushed down the drain.

⚠ *CAUTION*s warn about safety hazards.
*EXTRA*s give helpful hints, additional information, or interesting facts.

Reagents
6 M sulfuric acid [H_2SO_4]
butyric acid [$CH_3(CH_2)_2COOH$]
salicylic acid [$C_7H_6O_3$]
concentrated acetic acid [CH_3COOH]
formic acid [$HCOOH$]
propanoic acid [CH_3CH_2COOH]
methyl alcohol [CH_3OH] methanol
ethyl alcohol [CH_3CH_2OH] ethanol
benzyl alcohol [$C_6H_5CH_2OH$]
n-octyl alcohol [$CH_3(CH_2)_7OH$] n-octanol
isobutyl alcohol [$CH_3CH(CH_3)CH_2OH$]
n-pentyl alcohol [$CH_3(CH_2)_4OH$](n-amyl alcohol) n-pentanol

Laboratory Equipment
250 mL beaker
6 small test tubes
test tube holder
ring stand, ring, and wire gauze
laboratory burner

PROCEDURE

1. Prepare a hot-water bath by placing 150 mL of water in a 250 mL beaker over a Bunsen burner. Do not boil.

2. **Preparation of Methyl Salicylate**
   ("oil of wintergreen") Place a small scoop (0.1 g) of salicylic acid in a test tube. Note the aroma of salicylic acid. Add 15 drops of methyl alcohol and 3 drops of 6 M sulfuric acid to the test tube. Shake the tube to mix the reactants. Place the test tube in a hot-water bath. After a minute, note the aroma and appearance of the solution.

Salicyclic Acid   Methyl Alcohol

3. **Preparation of Octyl Acetate**
   ("orange peelings") Place 10 drops of concentrated acetic acid in a small test tube. Note the aroma. Add 10 drops of n-octyl alcohol and 2 drops of 6 M sulfuric acid. Shake to mix. Heat in the water bath and note the aroma and appearance of the solution.

Acetic Acid   n-Octyl Alcohol

4. **Preparation of Pentyl Acetate (Amyl Acetate)**
   ("banana oil") Place 15 drops of acetic acid, 15 drops of n-pentyl alcohol (n-amyl alcohol), and 3 drops of 6 M sulfuric acid in a small test tube. Shake to mix. Heat in the water bath and note the aroma and appearance of the solution.

Acetic Acid   n-Pentyl Alcohol

⚠ *CAUTION*
Never smell any chemical directly. Waft the vapor with your hand.
EXTRA
The different esters may require varying times to complete the reaction.
⚠ *CAUTION*
Sulfuric acid is corrosive. Don't point a test tube at anyone while heating.

EXTRA
If the aroma is not familiar, try diluting the ester with a bit of water to reduce its potency.

EXTRA
The chemical structure for each reagent is shown.

201

5. Preparation of Benzyl Acetate

("**jasmine**") Combine in a small test tube 20 drops of benzyl alcohol, 5 drops of acetic acid, and 3 drops of 6 M sulfuric acid. Shake to mix. Heat in the water bath and note the aroma and appearance of the solution.

$$
\begin{array}{cc}
& O \\
& || \\
H_3C-C-OH & \qquad HOCH_2-\text{\large\phi} \\
\text{Acetic Acid} & \qquad \text{Benzyl Alcohol}
\end{array}
$$

6. Preparation of Ethyl Propionate

("**rum**") Combine 20 drops of ethyl alcohol, 20 drops of propanoic acid, and 3 drops of 6 M sulfuric acid in a small test tube. Shake to mix. Heat the test tube in the water bath and note the aroma and appearance of the solution.

$$
\begin{array}{cc}
& O \\
& || \\
H_3CH_2C-C-OH & \qquad HOCH_2CH_3 \\
\text{Propanoic Acid} & \qquad \text{Ethyl Alcohol}
\end{array}
$$

7. Preparation of Isobutyl Formate

("**raspberry**") Combine 20 drops of formic acid, 20 drops of isobutyl alcohol, and 3 drops of 6 M sulfuric acid in a small test tube. Shake to mix. Heat the test tube in the water bath and note the aroma and appearance of the solution.

$$
\begin{array}{cc}
& CH_3 \\
O & | \\
|| & \\
H-C-OH & \qquad HOH_2C-C-CH_3 \\
& | \\
& H \\
\text{Formic Acid} & \qquad \text{Isobutyl alcohol}
\end{array}
$$

8. Preparation of Methyl Butyrate

("**apple**") Combine 20 drops of methyl alcohol and 3 drops of 6 M sulfuric acid in a small test tube. The rancid butter smell of butyric acid will easily be noted. In a hood, open the butyric acid, quickly add 5 drops to the test tube, and close the butyric acid container. Shake to mix. Heat the test tube in the water bath and note the aroma and appearance of the solution.

$$
\begin{array}{cc}
& O \\
& || \\
H_3CH_2CH_2C-C-OH & \qquad HOCH_3 \\
\text{Butyric Acid} & \qquad \text{Methyl Alcohol}
\end{array}
$$

EXTRA
Butyric acid is a real stinker! Do this reaction in a hood.

9. Preparation of Ethyl Butyrate

("pineapple") Combine 20 drops of ethyl alcohol and 3 drops of 6 M sulfuric acid in a small test tube. The rancid butter smell of butyric acid will easily be noted. In a hood, open the butyric acid, quickly add 5 drops to the test tube, and close the butyric acid container. Shake to mix. Heat the test tube in the water bath and note the aroma and appearance of the solution.

$$H_3CH_2CH_2C-\overset{\overset{\displaystyle O}{||}}{C}-OH \qquad HOCH_2CH_3$$

Butyric Acid   Ethyl Alcohol

EXTRA
Butyric acid is a real stinker! Do this reaction in a hood.

10. Preparation of Benzyl Butyrate

("cherry") Combine 20 drops of benzyl alcohol and 3 drops of 6 M sulfuric acid in a small test tube. The rancid butter smell of butyric acid will easily be noted. In a hood, open the butyric acid, quickly add 5 drops to the test tube and close the butyric acid container. Shake to mix. Heat the test tube in the water bath and note the aroma and appearance of the solution.

$$H_3CH_2CH_2C-\overset{\overset{\displaystyle O}{||}}{C}-OH \qquad HOCH_2-\langle\!\!\!\bigcirc\!\!\!\rangle$$

Butyric Acid   Benzyl Alcohol

EXTRA
Butyric acid is a real stinker! Do this reaction in a hood.

11. Each bottle of ester can be left open in a hood to evaporate unless otherwise directed by your instructor.

SYNTHESIS OF ESTERS
PRE-LAB QUESTIONS

Name: _____

Lab Partner: _____

Section: _____   Date:_____

1) Since we know that esters are produced from an alcohol and a carboxylic acid, would it be worth the time and expense for chemists to try every possible combination in an attempt to discover new scents and flavors? Support your opinion.

2) Any chemical process involves side reactions and a bit of unreacted reagent. Why should you not taste or use any of the ester you will produce in this lab?

3) How would the purity of the esters synthesized in this experiment compare with commercial products?

4) Vanilla flavoring is a combination of several compounds isolated in nature from the vanilla bean. However, only two of the compounds make up the largest percentage of the mixture and together those two compounds give the familiar flavor we know. Pure vanilla extract is quite expensive, but imitation vanilla, which contains the two major compounds synthesized in a lab, is not. List advantages and disadvantages for the use of the pure extract.

5) Many pharmaceuticals are synthesized copies of compounds found in nature. There is not enough of the natural substance to fill the needs of the world's population, and even if there were, the cost of isolating it from all the other materials that come with it is exorbitant. Why do you think that the widespread and inexpensive availability of compounds such as aspirin and insulin is a relatively recent phenomenon?

SYNTHESIS OF ESTERS
REPORT SHEET

Name: _____

Lab Partner: _____

Section: _____   Date: _____

I. For each compound, write a complete reaction equation showing the removal of the atoms that become the water molecule. Give observations of aromas and appearance. (One example is included in item 2.)

No need just ?'s

1. Methyl salicylate ("oil of wintergreen")

2. Octyl acetate ("orange peelings")

$$H_3C-\overset{\overset{\textstyle O}{\|}}{C}-OH \quad + \quad HO(CH_2)_7CH_3 \quad \underset{}{\overset{H^+}{\rightleftharpoons}} \quad H_3C-\overset{\overset{\textstyle O}{\|}}{C}-O(CH_2)_7CH_3 \quad + \quad H_2O$$

    Acetic Acid   Octyl Alcohol   Octyl Acetate   Water

    Acetic acid smells like strong vinegar and octyl alcohol smells like a combination of an antiseptic and a fuel, but the octyl acetate has a strong orange scent. All are clear liquids.

3. n-Amyl acetate ("banana oil")

4. Benzyl acetate ("jasmine")

5. Ethyl propionate ("rum")

6. Isobutyl formate ("raspberry")

7.   Methyl butyrate ("apple")

8.   Ethyl butyrate ("pineapple")

9.   Benzyl butyrate ("cherry")

II.   QUESTIONS:

1.   What are the physiological properties (effects on humans) of these esters? This information can be found in a <u>Merck Index</u>.

2.   Is it valid to be concerned about "synthetic" flavorings in food? Explain.

3.   List several common materials or products that contain esters.

4.   How would the use of esters change if they were not volatile?

26 Polymers

Megamolecules Out of Small Ones

OBJECTIVES

To produce a linear and a cross-linked polyester.

To compare properties of the two polyesters.

To prepare a take-home viscosity demonstration.

To prepare a nylon rope and calculate its length.

To distinguish from an equation whether a polymer is a mono- or a copolymer.

Relates to Chapter 10 of Chemistry for Changing Times, *twelfth ed.*

BACKGROUND

Polymers are large molecules produced by the combination of many small molecules called *monomers*. By changing monomers, chemists can produce different polymers. Chemists are often able to synthesize polymers with specific properties by choosing the right monomers.

Polymerization reactions are of two types. One type is the *addition reaction*, in which a double bond in each molecule is converted to a single bond as the molecules are joined together. Polyethylene is an example of this type of polymerization reaction. The second type is the *condensation reaction*, in which a small molecule such as water (H_2O), ammonia (NH_3), or hydrogen chloride (HCl) is produced along with the polymer bond. Polymers are also classified by the number of types of monomers that are used to make the polymer. A *monopolymer* is made of one type of monomer; a *copolymer* is made of more than one type of monomer.

One of the many areas in which polymers have had a major impact is in synthetic fibers. Many of the clothes that we wear are made from synthetic fibers. Two well-known synthetic fibers are nylon and polyester. There are also other uses besides clothing for these polymers.

Nylon is a polyamide because the bond formed between monomer units is an amide bond. Several different nylons are produced depending on the monomers used. The properties of the nylons also depend on the treatment after production. Nylons may be used in stockings after being made into thread, or molded into something like an unbreakable comb. One method of producing nylon is by reacting hexamethylenediamine and sebacoyl chloride. The nitrogen in hexamethylenediamine will bond to the carbonyl carbon (C=O), forming an amide bond. One hydrogen bonded to the nitrogen on the hexamethylenediamine and both of the chlorides bonded on the ends of the sebacoyl chloride are lost from their original molecules.

$$H_2N(CH_2)_6NH_2 \quad + \quad Cl-\overset{\overset{\displaystyle O}{||}}{C}-(CH_2)_8\cdot\overset{\overset{\displaystyle O}{||}}{C}-Cl \quad \longrightarrow \quad HCl \quad +$$

Hexamethylenediamine   Sebacoyl Chloride

$$-\overset{\overset{\displaystyle O}{||}}{C}-(CH_2)_8\cdot\overset{\overset{\displaystyle O}{||}}{C}-NH(CH_2)_6NH-\overset{\overset{\displaystyle O}{||}}{C}-(CH_2)_8\cdot\overset{\overset{\displaystyle O}{||}}{C}-NH(CH_2)_6NH-$$

Nylon 6,10

Because there are 6 carbons in hexamethylenediamine and 10 carbons in sebacoyl chloride, the nylon formed is called nylon 6,10. If adipoyl chloride (with 6 carbons) instead of sebacoyl chloride is used, the product is nylon 6,6.

Different products can be made from the same polymer. Polyester can be produced as a solid, or it can be drawn into a thread. Almost everyone has some clothing made of polyester. In this investigation you will produce a hard, solid polyester.

An example of making a polymer with specifically desired properties is the production of two different polyesters. They are called polyesters because of the ester (COC=O) bond between the monomers. In both esterification reactions the COC bond in phthalic anhydride is broken to form two new bonds with alcohol groups (OH). A linear polyester is formed when there are only two reactive sites (OH) on the alcohol, as with ethylene glycol.

Phthalic Anhydride   Ethylene Glycol

Linear Polyester

When glycerol, which contains three OH groups, is reacted instead of ethylene glycol, a cross-linked, three-dimensional polyester called glyptal is formed. Glyptal is a harder, more brittle compound than the linear polyester. Glyptal is used as a surface coating.

Phthalic Anhydride   Glycerol

Glyptal

WASTE AND THE ENVIRONMENT

Hexamethylene diamine, sebacoyl chloride, and hexane are all irritating to humans. Ethylene glycol is toxic. It is better not to place any of these in the water system. Concentrated acid can damage plumbing if not neutralized or diluted.

⚠ *CAUTION*s warn about safety hazards.
*EXTRA*s give helpful hints, additional information, or interesting facts.

Reagents

hexamethylenediamine [$H_2N(CH_2)_6NH_2$]
sodium hydroxide pellets [NaOH]
phenolphthalein (or food coloring)
sebacoyl chloride [$ClCO(CH_2)_8COCl$]
hexane [C_6H_{14}]
phthalic anhydride
sodium acetate
glycerol
ethylene glycol
sulfuric acid 18 M (for disposal)
Parafilm®
distilled water

Common Materials

large metal paper clips
aluminum foil (tear 4" strips from roll and halve)
food coloring (or phenolphthalein)

Laboratory Equipment

disposable test tubes
2 watch glasses
ring stand
2 utility clamps
tweezers or forceps
150 mL beaker
large beaker (400- or 600 mL)

PROCEDURE

1. Clamp two test tubes in a slanting position, close together on a ring stand. To each add 2.5 g of phthalic anhydride and 0.1 g of sodium acetate. Stir the solids to mix them thoroughly. To one of the test tubes add 1.0 mL of glycerol. To the other test tube add 1.0 mL of ethylene glycol. Point the tubes away from everyone and heat the test tubes gently. Heat each tube equally. The apparent boiling of the solutions is due to the escape of water produced as the condensation reaction occurs. Continue to heat after the "boiling" starts until the liquid just begins to yellow. Then remove the heat source.

2. While one student heats the reactants in the two test tubes, the other student should cover the inside of two watch glasses with aluminum foil.

3. When the reaction is complete, pour the contents of each test tube onto a foil covered watch glass while the contents are still hot. Allow the polymers to cool.

4. Test the two products for viscosity, elasticity, and brittleness.

5. Put concentrated sulfuric acid in the test tubes to dissolve the polyester in the test tubes.

6. Clamp a third disposable test tube in a slanting position on a ring stand. Add 2.5 g of phthalic anhydride, 0.1 g of sodium acetate, and 1.0 mL of glycerol. Point the tube away from everyone and heat the test tube gently. The apparent boiling of the solution is due to the escape of water produced as the condensation reaction occurs. Continue to heat until a light amber color first appears. Allow the product to cool in the test tube.

7. When it has cooled, stopper the test tube. Wrap the glass/stopper interface with Parafilm®. To make an interesting take-home display, open up a large paper clip and insert the test tube inside it to make a stand to prevent the test tube from rolling. The fluid will look like it is about to pour out of the test tube, but it will take a very long time to pour out.

⚠ CAUTION
Don't heat too long in one spot. Keep moving the flame.

EXTRA
Viscosity is the resistance to flow. The thicker the fluid, the more viscous.
⚠ CAUTION
Sulfuric acid is caustic.

⚠ CAUTION
Don't heat too long in one spot. Keep moving the flame.

8. During the time the polyesters are cooling, make a nylon rope. Prepare a 0.5 M solution of hexamethylenediamine by adding 0.5 g of sodium hydroxide (NaOH) and 1.5 g of hexamethylenediamine in 25 mL distilled water. Place 25 mL of 0.5 M hexamethylenediamine in a 150 mL beaker. Hexamethylenediamine can be dispensed by warming the reagent bottle in hot water. Decant the liquid when sufficient solid dissolves. You may add food color or phenolphthalein to the solution for color.

EXTRA
The pink color of the indicator may fade when the nylon is washed.

9. Prepare 0.2 M sebacoyl chloride by dissolving 1 mL of sebacoyl chloride in 25 mL of hexane. Slowly pour 25 mL of 0.2 M sebacoyl chloride down the side of the beaker, forming a second layer on top of the hexamethylenediamine solution. Disturb the hexamethylenediamine layer as little as possible.

▲ *CAUTION*
Hexane is flammable.

▲ *CAUTION*
Hexamethylenediamine and sebacoyl chloride are irritating to skin, eyes, and the respiratory system. Hexane can irritate the respiratory system.

10. Measure the diameter of a large beaker that you will use to wind the rope as it is produced. From this measurement, find the circumference.

11. With tweezers, carefully grasp the interface between the solutions and pull it out slowly and carefully. Do not jerk. Catch the end around the large beaker. As you rotate the large beaker, pulling more of the nylon out of the 150 mL beaker, the two solutions will continue to react at the interface until one of the solutions is consumed. Count the number of turns of the large beaker that are required to consume the limiting reactant.

EXTRA
Be sure to pull the nylon rope straight up out of the beaker so that it does not touch the side.

12. Thoroughly wash the nylon rope in water or ethanol before handling.

13. Calculate the length of the rope and test for viscosity, elasticity, and brittleness.

14. You may throw the solid polymers and foil into the trash. When all the polyester has dissolved, the sulfuric acid in the test tubes can be flushed down the drain with lots of water. The nylon reactants should have been used up. If not, follow the directions of your instructor as to their disposal.

POLYMERS
PRE-LAB QUESTIONS

Name: _____

Lab Partner: _____

Section: _____   Date: _____

1) Study the structural reaction equations for the production of the two polyester polymers. Are the polymer products monopolymers or copolymers?

2) What structural difference do you see in the molecules of the two polyesters? How might the structure alter the viscosity, mechanically?

3) Is the nylon a monopolymer or a copolymer?

4) Which reactions are condensation reactions and which are addition reactions?

POLYMERS
REPORT SHEET

Name: _____

Lab Partner: _____

Section: _____   Date: _____

I. INSTRUCTOR'S INITIALS

LINEAR: _____   GLYPTAL: _____   NYLON: _____

II. ROPE LENGTH
Circumference of large beaker _____cm $(C = \pi \times diameter)$

Number of turns of beaker _____

Rope length = circumference × turns = _____cm × _____

Rope length = _____cm

III. QUESTIONS
1. Which of the polyesters is more viscous?

2. Did the difference in viscosity fit your prediction based upon the structures of the molecules?

3. Is either polyester brittle?

4. Describe the properties of the nylon rope.

5. Did you find that the strength and brittleness of the nylon rope changes as it dried? Support your answer with your observations and theories.

6. Why was the experimental procedure designed so that there were only small amounts of the excess reactants remaining?

INVESTIGATION

27 *Polymer Properties*

Does It Bounce, Bend, or Tear?

OBJECTIVES

To observe how the treatment of a polymer during its formation affects its final properties.
To test polymers for several physical properties.
To determine how differences in reactant composition affect physical properties.

Relates to Chapter 10 of Chemistry for Changing Times, *twelfth ed.*

BACKGROUND

Polymer characteristics are due in part to the monomers that comprise them. Other factors that influence polymer properties are the types and number of bonds that form between polymer strands. Some polymers are composed of branched chains that also have an effect on the properties. In this investigation we will examine several polymers that are easily produced. We will vary the density and other properties by altering specific factors during their formation.

Polystyrene is a polymer made of one type of *monomer*, or small molecule, that links to itself repeatedly, forming a chain. *Styrofoam* is polystyrene into which a gas is blown as it forms. This produces a lightweight material with excellent insulating properties. Polystyrene without the gas blown into it (called *styrene*) is a very hard but brittle material used for radio cases and "fake" wood furniture. In this investigation you will have the opportunity to see how much difference there is in the densities of the two forms of polystyrene.

Polystyrene

Cross-linking polymer chains provides stability to the polymer. The polymer becomes more rigid but also more resilient because the extra cross-linking makes it return to its original shape. Polyvinyl alcohol and Elmer"s Glue® are polymers that can be cross-linked by sodium borate (Borax®). In the polyvinyl alcohol-borate polymer (often called "slime"), the boron (B) is bonded to four oxygens (O) that have lost hydrogens (H). These four oxygens can be attached to two polymer strands or four polymers strands, forming cross-links between the strands within the polymer. The more boron, the more the polymer is cross-linked. In this investigation you will have the opportunity to observe the results of the differences in boron content.

Polyvinyl Alcohol

219

$$\left(\begin{array}{c} H\ H\ H\ H\ H\ H \\ -C-C-C-C-C-C- \\ O\ H\ O\ H\ O\ H \end{array}\right)$$

Polyvinyl Alcohol
Crosslinked with Boron

Some of the properties that are affected by cross-linking are tensile strength, elasticity, density, resiliency, and "bounce." *Tensile strength* is measured by pulling quickly on the polymer. A high tensile strength will resist tearing. *Elasticity* determines how far the material will stretch and is tested by pulling on the material slowly. *Resiliency* determines how well it will return to its original form. "Bounce" measures how high it will bounce when dropped from a certain height.

WASTE AND THE ENVIRONMENT
The materials used in this investigation are nontoxic.

⚠ *CAUTION*s warn about safety hazards.
*EXTRA*s give helpful hints, additional information, or interesting facts.

Reagents
    acetone
    polyvinyl alcohol

Common Materials
    10 oz Styrofoam® cups
    paper clips
    4% Borax® (sodium borate) solution
    8% Borax® (sodium borate) solution
    Elmer"s Glue®
    12" rulers (marked in cm)

Laboratory Equipment
    100 mL graduated cylinder
    150 mL beaker
    stirring rod

PROCEDURE

1. **Styrofoam to Styrene**
   Determine the density of Styrofoam®. Weigh and record the mass of a 10-oz. Styrofoam® cup. Determine the volume by volume displacement. Fill a 100 mL graduated cylinder with water to 60 mL. Break the cup into pieces that will fit into the cylinder. Place the weighed pieces of Styrofoam® in the cylinder. Hold the Styrofoam® under the water with a straightened paper clip. Read the new volume. Find the volume of the piece by subtracting the original volume from the new volume. Density is mass divided by volume. Put the wet pieces of Styrofoam® into a trash can.

   EXTRA
   Water in the acetone will prevent it from dissolving the Styrofoam.

2. Place 10 mL of acetone in a 250 mL beaker. Break up three 10-oz. Styrofoam® cups into pieces. Add the pieces to the acetone one at a time until all of the pieces are dissolved. (When dissolving is complete, no more air bubbles are released.)

   EXTRA
   Acetone will remove your nail polish.
   ⚠ *CAUTION*
   Acetone can be absorbed by your skin. It would be safer to wear gloves.

3. Let the solution sit until all the acetone is absorbed. Remove the polymer from the beaker and knead it until it becomes firm, forms an outer skin, and becomes tough. The longer the polymer is kneaded, the more the acetone will evaporate. The polymer is now 'styrene." Set the polymer aside to dry.

   EXTRA
   Knead like bread. Push, pull, squeeze, twist, roll, and punch.

4. At the end of the laboratory session, determine the density of the hard styrene polymer. Weigh the polymer. Measure the volume by volume displacement. Hold it underwater with the straightened paper clip. Density is mass divided by volume.

   Which is more dense? Differences??

5. **Borate Polymer I**
   Dissolve 4 g polyvinyl alcohol in 100 mL of water. Warm and stir for several minutes. The solid is dissolved when the solution is clear.

6. While stirring constantly, slowly add 12 mL of 4% sodium borate ($Na_2[B_4O_5(OH)_4]\cdot8H_2O$) to the polyvinyl alcohol solution.

   EXTRA
   Borax® is sodium borate.

7. Pour the solution onto a clean, dry bench top. Roll and knead the material until it becomes a firm ball.

   EXTRA
   Try to keep it together in one piece. It does not "heal" well when separated.

8. **Borate Polymer II**
   Place 30 mL of Elmer's Glue® in a 150 mL beaker.

9. While stirring constantly, add 4% sodium borate until no more borate is absorbed. Form the polymer into a firm ball and then rinse it under running water.

10. **Borate Polymer III**
    Place 30 mL of Elmer's Glue® in a 150 mL beaker.

11. While stirring constantly, add 8% sodium borate until no more borate is absorbed. Form the polymer into a firm ball and then rinse it under running water.

12. Test polymers I, II, and III for bounce by dropping each ball beside a ruler from 30 cm. above the lab bench. Observe how high the ball bounces. Do several trials for each ball. Record the largest bounce for each ball.

EXTRA
It is important to determine the bounce height before you get the polymer wet.

13. Determine the density of each polymer. Determine the mass by weighing on a balance. Determine the polymer volume by volume displacement in a 100 mL graduated cylinder. Roll the polymer balls into cylinder shape with a diameter smaller than the inside of the graduated cylinder. Put 35 mL of water in the cylinder and then slide the polymer in carefully, avoiding any splashing. Density is mass divided by volume.

14. Test polymers I, II, and III for elasticity by *slowly* pulling on a piece of each polymer. Record your observations.

EXTRA
Dry each polymer before you try the elasticity or tensile strength tests.

15. Test polymers I, II, and III for tensile strength by *quickly* pulling on a piece of each polymer. Record your observations.

16. You may deposit the polymers in a trash container or take them home. Rinse the extra solutions down the drain.

POLYMER PROPERTIES
PRE-LAB QUESTIONS

Name: _____

Lab Partner: _____

Section: _____ Date: _____

1) Styrofoam is very brittle. What will you expect to find when you examine the hardness of styrene?

2) The slime produced in steps 5–7 will continue to cross-link and exclude water molecules from its matrix until all the water is gone. What will you expect to happen to the elasticity of the slime with time?

3) As water leaves the glue–borate polymers produced in steps 8–11, they become less and less elastic until, eventually, they are solids. How will you store them if you want to keep them?

4) Why might a toy manufacturer be interested in polymer properties?

POLYMER PROPERTIES
REPORT SHEET

Name: _____

Lab Partner: _____

Section: _____   Date: _____

X **STYROFOAM TO STYRENE**

	Mass	Initial Volume	Final Volume	Sample Volume
Styrofoam	_____ g	_____ mL	_____ mL	_____ mL
Styrene	_____ g	_____ mL	_____ mL	_____ mL

Density   Styrofoam: _____ g/mL   Styrene: _____ g/mL

II. **BORATE POLYMERS**   I   II   III

Highest bounce   _____ cm   _____ cm   _____ cm

	Mass	Initial Volume	Final Volume	Sample Volume
I	_____ g	_____ mL	_____ mL	_____ mL
II	_____ g	_____ mL	_____ mL	_____ mL
III	_____ g	_____ mL	_____ mL	_____ mL

Density   I: _____ g/mL   II: _____ g/mL   III: _____ g/mL

Elasticity and resiliency observations

I

II

III

✗ Differences between Styro + in Acetone

Tensile strength observations

I

II

III

III. QUESTIONS

1. Why are the densities of styrene and Styrofoam so different?

2. Which of the three borate polymers is the most dense?

3. Which of the three borate polymers has the most bounce?

4. Is there a relationship between density and bounce?

5. Which of the three borate polymers is the most elastic?

6. Is there a relationship between resilience and bounce? How about the relationship between resilience and the amount the ball deformed as it hit the surface?

7. Which of the three borate polymers has the most tensile strength?

28 *Nuclear Radiation*

Will You Glow in the Dark?

OBJECTIVES

To learn the use of a Geiger counter.

To be able to describe alpha, beta, and gamma rays.

To discover the presence of background radiation and evaluate the amount.

To demonstrate the depth of penetration of each type of radiation.

To measure the decrease of radiation with distance from the source.

To understand the use of radiation in a smoke detector.

Relates to Chapter 11 of Chemistry for Changing Times, *twelfth ed.*

BACKGROUND

The three most familiar types of nuclear radiation that are emitted by naturally occurring isotopes are alpha, beta, and gamma rays. The *alpha particle,* or ray, is a helium nucleus. It causes appreciable ionization as it travels through material; however, the path length for an alpha particle is very short. An alpha particle is usually stopped by a sheet of paper or a few centimeters of air.

The *beta particle,* or ray, is an electron. It will cause some ionization as it travels through material but much less than the alpha because it is much lighter and carries half of the charge. Penetration depth depends on the energy of the particle, but most beta particles are stopped by aluminum foil.

The *gamma ray* is simply energy emitted from the nucleus. A gamma ray will cause a small amount of ionization, but a high-energy ray (one with a short wavelength) will travel a long distance in material. Several feet of lead is required to stop the most energetic gamma rays.

A Geiger counter is a Geiger-Mueller tube attached to a box with meters on it. The *Geiger-Mueller tube* is a gas-filled shell with an electrode down the center. There is a small charge on the center electrode. When radiation goes through the tube, it ionizes the gas. The electrons are attracted to the central electrode, and the positive ions are attracted to the shell. The ion movement generates a potential difference between the shell and the center electrode. The electronics attached to the Geiger-Mueller tube use that potential difference to make a audible click or a needle movement on the meter. Alpha rays are not usually detected by Geiger counters, because they cannot get through the shell. Beta rays usually can be counted only if the tube is fitted with a thin window to let the beta rays through. Gamma rays are the usual radiation measured by a Geiger counter.

A radioactive material, or source, emits radiation in all directions. As the Geiger counter is moved farther away from the source, the tube will cover a smaller part of the sphere around the source. Also, some of the rays may be stopped by the air. Thus, as the Geiger counter is moved away from the source, the count rate will decrease.

Protection from radiation requires that the radiation be stopped by shielding or that the person be far enough away that only a small percentage strikes the person. A thin shield is all that is needed for alpha rays. Beta rays require a thickness of aluminum foil to stop most of the rays. Gamma rays require a lot of matter (several feet of dense material) to stop all of the rays.

One use of radiation is in the home smoke detector, which contains a small amount (1 microcurie [1 µCi], or about 0.32 µg) of americium-241. Am-241 emits an energetic alpha particle and low-energy gamma rays. A voltage difference applied to two plates forms an ionization chamber. The alpha particles ionize the air between the plates, causing a current to flow. When heavier particles like smoke enter the chamber, some of the alpha particles are blocked, the current is reduced, and the alarm sounds. Even steam will cause the alarm to sound, since the minute particles of water vapor are enough to stop some of the alpha particles. The smoke detector is safe because all the alpha particles are stopped by the case on the detector or by air. The gamma rays are of such low energy that the case also stops them. If the case was opened to expose the source, a Geiger counter would detect the gamma rays. The alpha rays would not penetrate the Geiger tube. Opening the case to expose the source is not recommended. Ingesting or inhaling the Am-241 could cause severe damage to the stomach or lungs.

Although radioactive mantles are difficult to find now, lantern mantles contained thorium for many years. Thorium produces a bright glow that was excellent for its use in a lantern. Radioactive radium was used to make luminous dials in watches for many years. Also, the orange pigment on older Fiestaware® dishes contains a radioactive uranium salt used for the color it adds to the pigment. Since the uranium is baked into the glaze coating on the dishes, the uranium atoms cannot transfer to other objects, but the dishes are a source of radiation.

About half the radiation humans are exposed to is natural radiation. Many soils contain uranium or thorium and their daughter isotopes, most of which are also radioactive. One of these daughter isotopes—radon, a noble gas—can collect in homes built over soils containing radioactive materials. Radon has become a health problem for people in those homes. Also, many common building materials made from the Earth's crust, such as bricks and concrete, are slightly radioactive. These and other sources of radiation are always present in the environment, as is a small amount of radiation from space. This ever-present level of radiation is known as *background radiation*. Humans have been living with it for all of their existence.

Radiation exposure is increased for smokers. Radioactive polonium-210 is inhaled by smokers. A person smoking two packs of cigarettes per day increases their radiation exposure to 20 times that of a nonsmoker.

WASTE AND THE ENVIRONMENT

Although there is plenty of natural radiation, it is important that extra radiation sources not contaminate the environment. If a source is to be disposed of, contact a trained person to facilitate the disposal.

▲ *CAUTION*s warn about safety hazards.
*EXTRA*s give helpful hints, additional information, or interesting facts.

Reagents
    alpha source
    beta source
    gamma source
    lead sheet

Common Materials
    several sheets of plain paper
    aluminum foil
    smoke detector

Laboratory Equipment
    Geiger counter

PROCEDURE

1. With no source to be tested in place, record the number of counts detected with the Geiger counter for 5 minutes. This number divided by 5 is the background minute count rate.

2. Place a beta source 1 cm from the Geiger counter. Measure and record the minute count rate. Repeat the count-rate determination at 1-cm intervals out to 10 cm.

3. Repeat step 2 using a gamma source.

4. Place an alpha source close to the Geiger counter to demonstrate that alpha rays are not detected.

5. With the beta source 1 cm from the counter, place a thin sheet of paper between the source and the counter. Record the minute count rate.

6. Add sheets of paper until the count rate is the same as the background count rate.

7. Replace the paper with a sheet of aluminum foil.

8. With the gamma source 5 cm from the counter, place a sheet of paper between the source and the counter. Record the minute count rate.

9. Repeat step 8 using aluminum foil and then a piece of lead or other heavy metal in place of the paper.

10. The instructor may take a count rate on the outside of a smoke detector. After opening the case, use the counter to identify the Am-241 source by finding the high gamma count rate.

NUCLEAR RADIATION
PRE-LAB QUESTIONS

Name: _____

Lab Partner: _____

Section: _____   Date: _____

1) Which of the three types of radiation is the most penetrating?

2) A certain percentage of all carbon in living systems is radioactive carbon-14. How does this relate to "background" radiation?

3) Is food served from older pieces of orange Fiestaware® radioactively contaminated? Would the food served on Fiestaware® emit anything that would register on a Geiger counter?

4) There are many uses of radiation in medicine. Radioactive sources can be used to tag the activity in specific organs of the body for diagnostic studies. Radiation is often used to alter the course or rate of progression of certain diseases. Why then are we concerned with adding radiation sources to the environment, or with increasing amounts of radiation that make it through our atmosphere?

NUCLEAR RADIATION
REPORT SHEET

Name: _____

Lab Partner: _____

Section: _____   Date: _____

I. BACKGROUND RADIATION

_____counts/5 min = _____ counts/min

II. Distance from Source

	1 cm	2 cm	3 cm	4 cm	5 cm	6 cm	7 cm	8 cm	9 cm	10 cm
BETA	____	____	____	____	____	____	____	____	____	____
GAMMA	____	____	____	____	____	____	____	____	____	____

III. ALPHA   _____

IV. BETA

No Paper	1 Sheet	2 Sheets	Foil
_____	_____	_____	_____

Number of sheets required to match background level _____

V. GAMMA

No Paper	1 Sheet	Foil	Lead
_____	_____	_____	_____

VI. SMOKE DETECTOR   Outside case _____   next to source _____

VII. QUESTIONS
1. Would you be safe from an alpha, beta, or gamma radiation source if you were a mile away?

2. Why are people not concerned about radiation sources in their homes, such as luminous clock faces, smoke detectors, indicator lights on appliances (krypton-85), electric blankets (promethium-147), and fluorescent lights (thorium-229)?

3. What would be the problem with using a radioactive isotope that emitted high-energy gamma rays in addition to the alpha particles in a smoke detector?

4. Why did manufacturers stop making luminescent watch hands with radium? (Radium emits high-energy gamma rays.)

5. If the polonium is in the tobacco, would being in the room with a smoker increase your radiation exposure?

6. Describe how the Geiger counter is designed to allow a person to differentiate between beta and gamma rays.

29 *Salt on Ice*

Why Does the Ice Melt Faster?

OBJECTIVES

To observe the lowering of the normal freezing point for a solvent as the concentration of dissolved ions increases.

To observe the effect that the number of ions in a formula has on the melting point of ice.

Relates to Chapter 12 of Chemistry for Changing Times, *twelfth ed.*

BACKGROUND

Sodium chloride, commonly called table salt or salt, occurs naturally in many places in the world. The two ions (sodium and chloride) are necessary for the proper functioning of our bodies. Salt is a common flavoring agent, and it is also used as a preservative since many bacteria do not survive in a salty environment.

Salt is also combined with ice in making homemade ice cream and is thrown on streets and sidewalks to melt ice. The purpose of using salt on icy streets is to make the ice melt at a lower temperature. If the melting point of the ice can be lowered, it might be low enough for the ice to melt and thus clear the street of ice. The presence of salt also tends to inhibit the formation of more ice. In making ice cream, we want the ice to absorb heat more quickly (make the ice cream that forms colder) so we lower the melting point. Melting the ice more quickly and absorbing heat more quickly are different perspectives of the same phenomenon.

The melting of ice is an equilibrium process. Ice is melting and water is freezing at the same time. The melting ice absorbs heat, the freezing water releases heat, and the surroundings stay close to the freezing temperature of water. If the overall effect is that the piece of ice is getting smaller, then ice is melting a little faster than water is freezing. When salt is added to the water, the freezing point of the water solution is lowered, so that ice is melting faster than water is freezing and the heat absorbed by the melting ice lowers the temperature of the ice-water mixture and the surroundings.

In the making of ice cream, the surroundings include the ice cream, which gets colder by releasing heat to the ice-water mixture. When salt is put on the ice on the street, the freezing point of the ice decreases and the temperature of the surroundings are now higher than the melting point of the ice, so the ice melts. Many locations use the salt calcium chloride, or magnesium chloride, instead of sodium chloride because they will lower the freezing point of water more than will sodium chloride. Because calcium chloride ($CaCl_2$) or magnesium chloride ($MgCl_2$) produce three ions in solution (one Ca^{2+} ion and two Cl^- ions) compared with two ions (Na^+ and Cl^-) from sodium chloride, there is a greater concentration of ions in the calcium chloride or magnesium chloride solution. The lowering of the freezing point of a solution depends on the concentration of ions, not on the identity of the ions.

How fast the ice melts or how fast it absorbs heat depends on the amount of salt added to the ice. The more salt, the faster the ice melts (the lower the freezing point) up to the point at which the water cannot dissolve any more salt.

WASTE AND THE ENVIRONMENT
None of the solutions contain toxic materials.

⚠ *CAUTION*s warn about safety hazards.
*EXTRA*s give helpful hints, additional information, or interesting facts.

Reagents
isopropyl alcohol [C_3H_7OH]
sodium chloride [$NaCl$]
calcium chloride [$CaCl_2$]

Common Materials
ice

Laboratory Equipment
1000 ml beaker
large test tube with a 1-hole stopper to fit
thermometer

PROCEDURE

1. Weigh a clean, dry 1000 mL beaker. Report the beaker mass in the report table. Add ice to the beaker up to the 600 mL line. Weigh the beaker and ice. Record the initial mass in the report table. Add tap water to the beaker up to the 600 mL line.

2. Put 25 mL of isopropyl alcohol in a test tube. Place a one-hole stopper in the test tube. Place a thermometer through the hole in the stopper. Run tap water over the test tube for 2 minutes. Using the thermometer, measure and record the temperature of the alcohol.

3. Submerge most of the test tube into the beaker of ice-water. Stir gently, keeping the alcohol level lower than the ice-water level but not allowing any water into the alcohol test tube.

4. After 1 minute, 2 minutes, 5 minutes, and 10 minutes, record the temperature of the alcohol. At 10 minutes, pour the water off the ice, wipe any frost or condensation from the beaker, and weigh the beaker and ice. Record the final mass in the report table.

5. Warm the alcohol to its original temperature by running tap water over it or by setting it in a beaker of tap water or warm water while preparing for the next step.

6. Prepare the ice beaker as before using the same mass of ice, except add 10 g of sodium chloride (NaCl) to the new ice-water mixture. When the alcohol, thermometer, and test tube are at the original temperature, repeat steps 3 and 4.

7. Repeat steps 5 and 6 for 30, 50, and 70 g of sodium chloride, making sure the alcohol, thermometer, and test tube temperature at the beginning of each run is the original temperature and that the ice-water mixture is prepared fresh each time.

8. Repeat steps 5 and 6 for 10 g of calcium chloride ($CaCl_2$), making sure the alcohol temperature at the beginning of the run is the original temperature and that the ice-water mixture is prepared fresh.

9. You may flush the solutions down the drain with plenty of water.

⚠ CAUTION
Be careful when inserting the thermometer. It will help to use a lubricant such as petroleum jelly or glycerin.

⚠ CAUTION
A frosted beaker can freeze to your hand.

⚠ CAUTION
The test tube can break if warmed too quickly.

EXTRA
This is a cool lab.

SALT ON ICE
PRE-LAB QUESTIONS

Name: _____

Lab Partner: _____

Section: _____   Date: _____

1) Salt produces ions when it dissolves in water. However, the ions also provide the means for faster rusting of most iron-containing substances. What are the implications of this practice for automobile owners in areas that have many days of freezing weather?

2) There are salts that would deliver more ions in solution than magnesium or calcium chloride. Aluminum sulfate would deliver five ions per formula unit, but the sulfate is an ion that could alter the pH of the water in the environment. What are some other effects that would have to be taken into consideration before choosing a salt for the icy roads?

3) Can you think of a reason why people don't use calcium chloride or magnesium chloride in their ice-cream freezers at home?

4) The salt that is used to melt ice on roads re-crystallizes once the ice is melted and the water has evaporated. The trouble is that most if it is on the underside of cars or on the roadside, where all the slush was thrown by cars passing cars. Think of an alternative to ice that would increase safety in cold weather, but not spread foreign ions in the environment. Would it work as well for continual freezing rains?

SALT ON ICE
REPORT SHEET

Name: _____

Lab Partner: _____

Section: _____   Date: _____

I.

Salt Added	0 g NaCl	10 g NaCl	30 g NaCl	50 g NaCl	70 g NaCl	10 g CaCl$_2$
Original temperature						
Temperature at: 1 min.						
2 min.						
5 min.						
10 min.						
Final difference in temperature						
Initial mass of beaker + ice						
minus mass of beaker						
Initial mass of ice						
Final mass of beaker and ice						
minus mass of beaker						
Final mass of ice						
% ice melted= 100% × initial mass - final mass initial mass						

II. QUESTIONS

1. Was there a difference in the rate of cooling for the different amounts of salt?

2. Was there a difference in the amount of frost produced on the outside of the large beaker during each test run?

3. Is it possible for the surroundings to be so cold that salt will not cause the ice to melt?

241

4. Which amount of salt caused the most ice to melt?

5. Which amount of salt caused the lowest temperature?

6. Is it possible to add more salt than is needed?

7. Which caused the ice to melt faster, NaCl or $CaCl_2$? Why?

30 *Copper Cycle*

From Copper to ... Copper

OBJECTIVES

To produce and observe the chemical reactions involved in the transformation of copper metal to a series of four copper compounds and then recover the copper as a metal.

To classify reactions based on the observed reactants and products.

To write and balance a chemical equation for each reaction in the cycle.

To demonstrate the Law of Conservation of Mass.

To further develop lab skills of decantation, filtration, and precipitation.

To learn the proper use and operation of a Buchner funnel, dessicator, and analytical balance.

To use stoichiometry to predict the amount of a compound formed.

To calculate percent yield on the final copper product.

Relates to Chapters 12 and 5 of Chemistry for Changing Times, *twelfth ed.*

BACKGROUND

Relatively few metals are found in nature in their pure, elemental form. Most metallic elements appear naturally in a variety of substances, mainly as oxides and carbonates. Many of these metallic compounds have characteristic colors that can be traced to the specific metal content, and thus, colors can be used as a form of identification for many metallic substances.

Metals that are reactive are never found as pure elements in nature. These metals are always found bonded to at least one other element in the form of a compound. Such metals are found at the top of what is known as an *activity series*. The activity series lists metals in descending order of reactivity, or ease with which they react, with other elements and substances in the environment such as oxygen, water, and acids. Metals at the lower end of the activity series are ones that are not very reactive and can be found in the natural world in their elemental form. Examples of this type of metal are gold, platinum, silver, and copper. Because they are fairly un-reactive and do not easily corrode, they are used in coins and jewelry and are sometimes referred to as "precious" metals.

A great variety of compounds can be formed from most metals. A single metal can often be converted from one compound to another by a series of reactions. Each reaction requires the addition of a reactant and will produce at least one by-product. Some by-products are toxic or environmentally harmful substances that must be dealt with further, involving additional chemical processing. The relative cost of necessary reactants and the cost of disposal of unused by-products are major factors in the expense of industrial processes. In recent years, much has been said concerning the alternatives to mining new materials. The ability to recover metals from various substances is an important part of recycling programs that are designed to reduce the need for new sources of metals and continued mining operations. However, these processes may or may not always be the most cost-effective or the most environmentally friendly solution.

Many compounds are *soluble* in water; that is to say that they will separate into individual particles and become thoroughly mixed so that they are invisible to the naked eye. The resulting solution may or may not have a color to it that is indicative of the dissolved substance. Some reactions result in soluble compounds, while others result in *insoluble* compounds that show up as visible solid or semi-solid substances in the liquid. Insoluble substances that form in a solution are called *precipitates*. Insoluble substances can often be separated from a liquid by filtering or by *decanting*. Decanting works well for precipitates that readily settle to the bottom of the mixture. The liquid above the solid is called the *supernate*. Decantation is the process of pouring off the supernatant liquid, leaving the precipitate behind in the container.

Chemical changes may be indicated by a color change, gas formation, precipitation, and/or spontaneous energy changes. However, the only proof of a chemical change is the testing of the products to be certain that there are actually new substances with new properties that are distinctly different from the reactants with which you started.

In the course of chemical changes, there are often energy changes that can be detected by feeling of the reaction vessel or by using a thermometer to record the movement of heat into or out of the reaction system. These exchanges of heat or other forms of energy between the reaction system and the environment are referred to as *exothermic* if the system is losing heat (energy) to the environment, or *endothermic* if the environment is supplying heat (energy) to the system. If no energy is exchanged with the environment, we refer to the reaction as *isothermic*.

Beginning with pure copper nitrate, this experimental procedure will move the ionized copper through four different compounds and then finally recover the pure copper. Topics related to the mining, purification, recovery, recycling, and reuse of metals are illustrated through the exercise and have bearing on industrial as well as environmental issues.

After each compound is produced, you will identify the substances involved as reactants or products, soluble or insoluble, and the reaction as endothermic or exothermic.

In this experiment, copper metal will be dissolved by nitric acid to form copper nitrate, reacted with sodium hydroxide to form copper(II) hydroxide gel, heated to form solid copper(II) oxide, filtered and then reacted with hydrochloric acid to form aqueous copper(II) chloride, and finally reacted with aluminum to form copper metal. The reaction series can be summed up as follows:

$$Cu \rightarrow Cu(NO_3)_2 \rightarrow Cu(OH)_2 \rightarrow CuO \rightarrow CuCl_2 \rightarrow Cu$$

In each step, the copper is the limiting reactant and all other reagents added to the reaction sequence are designed to be in excess. In this manner, the most complete reaction with the copper can be encouraged. At the conclusion of each step, determine to the best of your ability that all the copper or copper compound has been reacted.

Anytime the reaction product containing copper is transferred to another reaction vessel, be certain to rinse all traces of the solid or solution into the new vessel, including the rinsing of stirring rods, forceps, and any other equipment that has come into contact with the substance. Do NOT place any lab equipment on any surface where it could transfer a copper-containing material.

WASTE AND THE ENVIRONMENT

Concentrated acids or bases can damage plumbing if they are not diluted and/or neutralized before disposal. All by-product solutions from this investigation will be near neutral and will contain no harmful substances. Everything except unused concentrated acid or base reagents and the final copper product can be flushed down the drain with plenty of water.

⚠ *CAUTION*s warn about safety hazards.
*EXTRA*s give helpful hints, additional information, or interesting facts.

Reagents
  clean copper turnings or wire
  red and blue litmus paper
  nitric acid, concentrated [HNO_3]
  8.0 M sodium hydroxide [NaOH]
  1 M monobasic sodium phosphate [Na_3PO_4]
  2 M sulfuric acid [H_2SO_4]
  6.0 M hydrochloric acid [HCl]
  filter paper circles to fit Buchner funnel
  mossy zinc
  distilled water

Laboratory Equipment
  10 mL graduated cylinder
  9.0 mm Buchner funnel with filter flask
  150 mL beaker
  250 mL beaker
  ring stand, ring, and wire gauze
  laboratory burner
  analytical balance
  forceps
  dropper

PROCEDURE

Part A: Preparation of Copper (II) Nitrate

1. Obtain about 0.5 g copper wire or copper turnings. (If you are using copper wire, you will need to use the following data to determine the length you need. An 18-gauge Cu wire is 0.102 cm in diameter and Cu density is 8.92 g/cm. Sand the wire to get all impurities off the wire.) Measure and record the exact mass using an analytical balance.

2. Place the copper metal into a 250 mL beaker and, working under a fume hood, measure 3 mL of concentrated nitric acid into a clean 50 mL beaker. Remaining under the fume hood, carefully add about 2.5 mL of concentrated HNO_3 one drop at a time to the copper in the larger beaker. Wait 1 minute and then stir gently if all the copper has not reacted. If solid copper remains, add another drop of concentrated HNO_3 and stir gently. Make as many observations as possible. Record these in your lab record. This procedure results in the formation of a ternary compound by the direct oxidation of a metal. Water is a by-product as well as a brown gas, NO_2. Before proceeding, check the solution by peering up through the bottom of the beaker to be certain no solid copper remains.

3. Dilute the $Cu(NO_3)_2$ solution with 10 mL of distilled water. Save this solution for Part B.

Part B: Preparation of Copper (II) Hydroxide

4. Obtain 5 mL of 8 M NaOH solution in a clean 50 mL beaker.

5. *Slowly* add nearly all of the NaOH to the $Cu(NO_3)_2$ solution, one drop at a time. Gently swirl the solution and then allow the precipitate to settle, forming a clear layer on top of the solid.

6. Add one more drop to the clear solution at the top, watching from the side of the beaker to see if any more precipitate forms.

7. When no new precipitate forms, dip the tip of the stirring rod into the clear portion at the top of the solution and then touch it to a piece of red litmus paper to be certain the solution is basic. (Litmus is red in pH<7 and blue in pH>7.) Stirring or strongly agitating the precipitate can cause it to change from $Cu(OH)_2$ to CuOH, which is a finer powder and a lighter aqua color. The CuOH will not be converted to CuO in the next step. Be careful not to transfer any precipitate to the litmus paper. Make as many observations as possible. Record these in your lab record along with a balanced equation and the classification of the reaction.

⚠ *CAUTION*
Copper turnings have sharp edges and can cause cuts.

⚠ *CAUTION*
Nitric acid is corrosive.

⚠ *CAUTION*
Remember that all colored gases are poisonous!

EXTRA
Keep the nitric acid volume to a minimum. More time to dissolve the copper now means less acid to be neu-tralized later and a better return on the copper in the long run.

⚠*CAUTION*
Sodium hydroxide solution is basic and is called an alkali *solution. It is caustic.*

EXTRA
The dilution of the sodium hydroxide is exothermic. Addition of the NaOH too quickly will heat the solution and bring on pre-mature conversion to copper oxide.

EXTRA
If the solution is basic, it is a good indication that the hydroxide ion has been bonded to all available copper ions with excess hydroxide ions left in solution and, thus, the reaction is complete.

Part C: Preparation of Copper (II) Oxide

8. Add enough distilled water to the beaker to bring the volume to 100 mL.

9. Boil the solution gently for 5 minutes, stirring constantly. The $Cu(OH)_2$ will decompose to give CuO. Check the precipitate while stirring to be certain no more blue color remains.

10. Set up a Buchner filtration apparatus. Wet the filter paper with distilled water. Filter the solution and rinse the last traces of the solid from the beaker and stirring rod with distilled water. Discard the clear, colorless liquid and flush it with plenty of water.

⚠ **CAUTION**
Use care when boiling a solution. Heat it slowly and do not allow it to spatter or "bump," as it can cause serious burns and result in a loss of material.

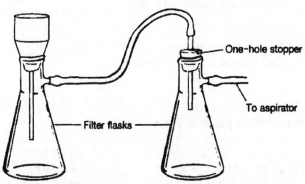

One-hole stopper

To aspirator

Filter flasks

Buchner funnel system with safety flask

11. Wash the CuO with hot water once to remove any other substances. Leave the filter paper containing the solid in the funnel for Part D. Water is a by-product of this reaction. Make as many observations as possible using all of your senses except taste. Record these in your lab record along with a balanced equation and the classification of the reaction.

EXTRA
The Buchner funnel can easily pull the filtrate into the water system if the level of filtrate gets too high in the filter flask. Keep an eye on it so that the water system is not contaminated.

Part D: Preparation of Copper (II) Chloride

12. Obtain 8 mL of 6 M HCl in a clean 50 mL beaker. Using a clean eye dropper, dribble the acid over the solid on the filter paper, allowing it to drain into a clean 250 mL beaker. Use the acid repeatedly until all the solid is reacted. If the reaction stops, add 1-mL 6 M HCl.

⚠ **CAUTION**
The acid is corrosive.

13. When all of the solid has been dissolved, the dropper, filter paper, and funnel should be rinsed thoroughly into the 250 mL beaker containing the $CuCl_2$ using distilled water. Be certain that all color is rinsed from the paper and all parts of the Buchner funnel. Save the solution for use in Part E. The reaction also produces water as a by-product. Make as many observations as possible using all of your senses except taste. Record these in your lab record along with a balanced equation and the classification of the reaction.

EXTRA
Be sure to separate the two pieces of the funnel to rinse the insides and all surfaces, decreasing the chance of material loss.

Part E: Preparation of Copper Metal

14. Measure about 1 gram of mossy zinc. Gently drop the zinc into the solution of $CuSO_4$. Allow it to stand, with occasional slow stirring, until the solution is completely colorless. Once the solution has cleared, remove the remaining zinc and rinse any loose copper from it and back into the beaker.

15. Allow the precipitate to settle and then decant and discard the supernatant liquid, flushing it with plenty of water. Note that the zinc will continue to react with the acid even after the copper has all been precipitated from the solution. It is important to remove the excess zinc when no more copper is being deposited. Solid zinc phosphate will cause error in your final mass. Make as many observations as possible. Record these in your lab record along with a balanced equation and the classification of the reaction.

16. Add 20 mL of cold distilled water to the copper in the beaker to wash it. Decant and discard the liquid. Repeat this two more times to wash the copper.

17. Record the mass of a clean 50 mL beaker taken on an analytical balance. Also, record the mass of a piece of filter paper for the Buchner funnel.

18. Assemble the filtering system and rinse the copper onto the filter paper. Continue to draw air through the filter for 5 minutes to dry the copper.

19. Weigh the beaker, paper, and copper. Determine the mass of the copper by difference and then calculate percent return on the copper metal.

20. The copper metal should be deposited in the trash at the conclusion of the lab.

EXTRA
Be careful not to lose any copper in the decantation process. It is better to leave a small amount of supernatant liquid in the beaker than to wash copper down the drain.

COPPER CYCLE
PRE-LAB QUESTIONS

Name: _____

Lab Partner: _____

Section: _____   Date: _____

1) In what instances would continued mining of a metal be preferable to recycling processes?

2) List one or two metals that would not be readily available for recycling programs because of the general manner in which they are used. Are these metals occasionally found in their pure form in nature? Are they relatively expensive or inexpensive metals?

3) Would you expect metals on the upper or lower end of the activity series to be more easily recyclable? Explain your reasoning.

4) When landfills are to be used for the disposal of water-soluble materials, what are the major environmental concerns?

5) Discuss the advantages and disadvantages of recycling efforts aimed at toxic metals.

6) How would you expect percent yield to be affected by increasing the number of reactions in a metal cycle? How would this most likely impact recycling processes?

COPPER CYCLE
REPORT SHEET

Name: _____

Lab Partner: _____

Section: _____   Date: _____

RESULTS

Part A. Preparation of Copper Nitrate

Mass of copper metal used   _____g

Maximum amount of copper that could possibly be recovered   _____g

Observations

Balanced equation

Part B. Preparation of Copper Hydroxide

Observations

Balanced equation

Part C. Preparation of Copper Oxide

Observations

Balanced equation

Part D. Preparation of Copper Chloride

Observations

Balanced equation

Part E. Preparation of Copper Metal

Observations

Balanced equation

Amount of copper recovered: _____g

Percent yield of copper recovered   _____%

QUESTIONS

1) If a student begins with 1.8 grams of copper metal and has a 67% return for the laboratory procedure, what mass of copper is present after drying?

2) Thinking of the implications in this procedure for the recycling of copper metals:
   a) What might be some drawbacks to recycling processes?

   b) What advantages are there to recycling copper?

   c) What environmental concerns come into question with the procedure?

   d) When would recycling not be the best way to deal with waste metals?

3) If a student begins with 0.85 grams of copper metal and expects 98% return, what is the minimum mass of zinc metal he should place into the final reaction mixture?

4) A student gets a return of 100.5%. Give as many plausible explanations as you can think of for the obvious error.

31 O_2 and CO_2 in Breath and Air

Gases, Gases Everywhere

OBJECTIVES

To determine by volume difference the percentage of carbon dioxide in exhaled air.

To determine by volume difference the percentage of oxygen in exhaled air and in normal air.

To illustrate the reaction of a gas with components of a solution.

Relates to Chapters 13 and 16 of Chemistry for Changing Times, *twelfth ed.*

BACKGROUND

The air around us is composed of nitrogen (N_2), oxygen (O_2), argon (Ar), carbon dioxide (CO_2), water vapor, and several trace gases. The argon, carbon dioxide, and trace gases combined account for just less than 1%. The amount of water vapor in air usually varies between zero and 5%. Unreactive nitrogen is the most common gas, at 78%.

Combustion reactions consume oxygen and produce carbon dioxide. The reaction product is still carbon dioxide whether in a flame or in our bodies.

$$\text{Hydrocarbon} + O_2 \rightarrow CO_2 + H_2O$$

The basis for this investigation is the reaction of carbon dioxide with calcium hydroxide to form a solid, calcium carbonate. By removing the gaseous carbon dioxide, we can measure the volume percent of carbon dioxide in a gas. The oxygen volume is measured by changing oxygen to carbon dioxide and measuring the volume of the carbon dioxide.

A candle inside a beaker will use up the oxygen in the trapped air and produce carbon dioxide as a product of the combustion reaction. All the oxygen atoms will be in the form of carbon dioxide when the flame goes out. The carbon dioxide will react with calcium hydroxide in the solution to produce calcium carbonate, but it requires some time for the carbon dioxide gas molecules to interact with the calcium ions and hydroxide ions in the solution. Carbon dioxide reacts with water to form carbonic acid (H_2CO_3). The two acidic hydrogen ions react with hydroxide ions to form water, leaving the carbonate ions to react with the calcium ion to form solid calcium carbonate. Pushing the beaker down increases the pressure on the gas and causes this reaction to occur more rapidly.

$$CO_2 + Ca(OH)_2 \rightarrow CaCO_3 + H_2O \text{ (net molecular equation)}$$

The difference in volume between that of the full beaker (which contained oxygen and other gases found in air) and the volume after the carbon dioxide has reacted (which contains only the other air gases) is the volume of oxygen in the air.

Some carbon dioxide and some oxygen will dissolve in water. Blowing into the water causes the water to become saturated with carbon dioxide and oxygen. When the air blown into the 100 mL flask is pushed into the 100 mL graduated cylinder, none of it will dissolve in the water. When the water is replaced with calcium hydroxide solution, the carbon dioxide in the exhaled air reacts with the calcium hydroxide. The resulting air trapped in the 100 mL cylinder is free of carbon dioxide. The two volumes can be subtracted to determine the volume of carbon dioxide, and from that the percentage of carbon dioxide.

During normal breathing, not all the oxygen inhaled into the body is consumed; part of it is exhaled with the carbon dioxide that the body expels. Displacing the exhaled breath from the flask by water into a beaker where a candle is burning causes the oxygen in breath to be used in a combustion reaction. The product of the reaction is carbon dioxide, which is absorbed by the solution of calcium hydroxide. The remaining volume of gas is the nitrogen from air and from breath. The volume of oxygen in breath can be calculated by subtracting the volume of nitrogen and the volume of carbon dioxide in breath. When the procedure is repeated after the candle is out, the volume becomes the sum of nitrogen in air and nitrogen and oxygen in breath. The difference between the results of the two procedures is the volume of oxygen in breath.

WASTE AND THE ENVIRONMENT

None of the solutions are harmful to the environment.

⚠ *CAUTION*s warn about safety hazards.

*EXTRA*s give helpful hints, additional information, or interesting facts.

Reagents
  calcium hydroxide solution, saturated
    [$Ca(OH)_2$]
Common Materials
  small birthday candles
  heavy book
  straws

Laboratory Equipment
  100 mL Erlenmeyer flask
  medium 1-hole rubber stopper (#5 or #6)
  3 lengths of tubing
  2 hose clamps
  2 1000 mL beakers
  wax pencil
  2-hole stopper to fit the Erlenmeyer flask
  250 mL beaker
  100 mL graduated cylinder
  funnel

PROCEDURE
Part A: Oxygen in Air

1. Place a small candle in a hole in the small end of a medium rubber stopper. It may be necessary to use hot wax to fit the candle securely to the stopper.

2. Place the stopper in a 1000 mL beaker. Place enough saturated calcium hydroxide, $Ca(OH)_2$, solution in the beaker to leave a half inch of the candle above the solution.

⚠ *CAUTION*
Calcium hydroxide is caustic.

3. Light the candle. Cover the candle by holding an inverted 250 mL beaker so that the lip of the beaker is just under the surface of the water.

4. Hold the inverted beaker in place until the flame goes out. Add saturated calcium hydroxide solution to the large beaker up to the 800 mL level while holding the bottom of the inverted beaker level with the top of the large beaker.

5. Set a heavy book on the top of the large beaker to keep the inverted beaker in place. Set the beaker aside to allow the air in the inverted beaker to cool and the carbon dioxide to react with the calcium hydroxide.

Part B: CO$_2$ in Breath

6. Fill a 1000 mL beaker with 900 mL of water. Insert a piece of thoroughly cleaned tubing or a straw into the water. Blow into the tubing for 2 minutes to saturate the water with carbon dioxide and oxygen.

7. Fill a 100 mL Erlenmeyer flask with water. Hold your hand tightly over the top and quickly invert the flask into the 1000 mL beaker so that the mouth of the flask is underwater and the flask remains full of water. Stick the end of the tubing into the flask. Blow through the clean tubing (straw) so that the flask fills with air.

8. Prepare the apparatus as shown below by completing steps 8 through 10. Close both clamps on the hose arrangement. Keeping the lip of the 100 mL Erlenmeyer flask just underwater, place the stopper of the hose arrangement securely into the 100 mL flask. Set the flask upright and set it aside.

9. Fill a 100 mL graduated cylinder with the breath-saturated water prepared in step 6. Cover the top tightly with your hand and invert it into the 1000 mL beaker so that the mouth of the cylinder is underwater and the cylinder is full of water. Clamp it in place using a ring stand and clamp.

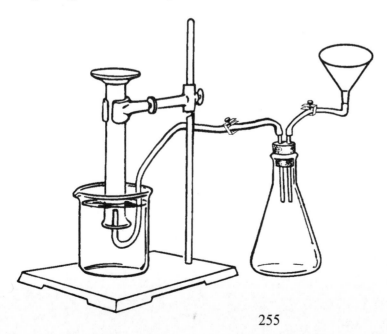

10. Insert the end of the hose arrangement into the 100 mL graduated cylinder as shown in the figure. Fill the funnel full of water. Slightly open both clamps. Add water to the funnel until the 100 mL flask is full of water.

11. Unclamp the cylinder. Hold it so that the water level inside is even with the water level on the outside. Read and record the level of the water. Remember that the cylinder is upside down.

12. Repeat steps 7 through 11 using saturated calcium hydroxide in the 100 mL cylinder. After inverting and clamping the cylinder, wash your hands.

⚠ *CAUTION*
Calcium hydroxide is caustic.

13. Carefully remove the book from the top of the 1000 mL beaker. Hold the inverted beaker so that the water inside is level with the water outside. With a wax pencil, mark the water level on the inverted beaker.

14. Remove the 250 mL beaker from the 1000 mL beaker. Set it upright. Using a 100 mL graduated cylinder, measure the amount of water needed to fill the beaker to the marked line and then to fill the beaker completely. Record the volumes on the report sheet.

Part C: Oxygen in Breath

15. Repeat steps 7 and 8. Insert the end of the hose arrangement under the edge of the candle beaker. Let the candle burn out. Fill the funnel full of water. Slightly open both clamps. Add water to the funnel until no more gas can be pushed out of the flask. Hold the 250 mL beaker so that the water inside is level with the water outside. With a wax pencil, mark the water level on the 250 mL beaker.

16. Repeat step 15, but add the water to the funnel while the candle is still burning.

17. Repeat step 14 for the two new marks on the beaker.

18. Decant the solutions from the settled solid and flush them down the drain with plenty of water. Throw the solids in the trash.

O_2 AND CO_2 IN BREATH AND AIR
PRE-LAB QUESTIONS

Name: _____

Lab Partner: _____

Section: _____   Date: _____

1) If we exhale mostly carbon dioxide, why does a person doing CPR blow into the mouth of an accident victim?

2) What common foods contain carbonic acid?

3) Does the fact that fish live in water prove that oxygen dissolves in water?

O_2 AND CO_2 IN BREATH AND AIR
REPORT SHEET

Name: _____

Lab Partner: _____

Section: _____   Date: _____

I. O_2 IN AIR

Volume of 250 mL beaker	_____	mL	Step 14
Minus final volume	− _____	mL	Step 14
Volume of O_2 in air	_____	mL	

$$\% \ O_2 \text{ in air} \ = \ \frac{O_2 \text{ volume} \times 100\%}{\text{Beaker volume}}$$

$$= \ \frac{\text{mL} \times 100\%}{\text{mL}} \ = \ \underline{\hspace{2cm}} \ \%$$

II. CO_2 IN BREATH

Volume of displaced H_2O	_____	mL	Step 11
Minus volume through $Ca(OH)_2$	− _____	mL	Step 12
Volume of CO_2 in breath	_____	mL	

$$\% \ CO_2 \text{ in air} \ = \ \frac{CO_2 \text{ volume} \times 100\%}{\text{Volume of displaced} \ H_2O}$$

$$= \ \frac{\text{mL} \times 100\%}{\text{mL}} \ = \ \underline{\hspace{2cm}} \ \%$$

III. O_2 IN BREATH

Volume of beaker with unlighted candle	_____	mL	Step 15
Minus volume of beaker with lighted candle	− _____	mL	Step 16
Volume of O_2 in breath	_____	mL	

$$\% \ O_2 \text{ in breath} \ = \ \frac{O_2 \text{ volume} \times 100\%}{\text{Volume of displaced} \ H_2O}$$

$$= \ \frac{\text{mL} \times 100\%}{\text{mL}} \ = \ \underline{\hspace{2cm}} \ \%$$

IV. QUESTIONS

1. What would be the effect on your results if the CO_2 did not come in contact with the $Ca(OH)_2$ solution?

2. What difference would you find if you pumped air from an air compressor instead of blowing air from your lungs in order to saturate the water?

3. What difference would you find if you pumped air from an air compressor instead of blowing air from your lungs to fill the flask?

4. If you used a compound other than calcium hydroxide, what would be the necessary requirement for that compound?

5. What would be the difference if calcium hydroxide were not a base?

6. Would CPR work if your exhaled breath had no oxygen in it?

7. How does this investigation relate to buffers in the blood and the optimum pH of blood?

32 *Solids in Smoke*

Smoke Gets in Your Eyes ... and That's Not All!

OBJECTIVES

To use the process of mass difference to make quantitative measurements of the solids produced by a burning cigarette.

To determine whether the smoker or a person exposed to secondhand smoke captures the largest amount of solids from smoke.

To make a quantitative determination of the segment of a cigarette that delivers the least mass of solid particles per mass of cigarette consumed.

Relates to Chapters 13 and 19 of Chemistry for Changing Times, *twelfth ed.*

BACKGROUND

Cigarette smoke is generated by burning tobacco leaves. The smoke produced in this combustion reaction has at least two major components that are harmful to the human body.

Carbon monoxide (CO), one of the major components of cigarette smoke, is an odorless and invisible but toxic gas. Carbon monoxide attaches to hemoglobin (making carboxyhemoglobin) and prevents hemoglobin (as oxyhemoglobin) from carrying oxygen to the rest of the body. The affinity of hemoglobin for carbon monoxide is 200 times greater than for oxygen. A cigarette smoker often has a carboxyhemoglobin (COHb) level of 5% instead of the normal 0.5%. An increased level of COHb can cause increased stress on the heart, impaired time discrimination, and high blood pressure.

Another harmful component in cigarette smoke is small solid particles. These particles may be fly ash, organic tar, or mineral dust. Some of these solids have been found to be carcinogenic or cancer-causing. The size of these particles can be as small as a few micrometers (10^{-6} m) or less. When solid particles enter the lung, they can irritate and destroy alveoli (tiny sacs where gases are exchanged) and cause emphysema.

The surgeon general of the United States has published research indicating that smoking is linked to lung cancer. A smoker may have ten times the chance of developing lung cancer as a nonsmoker. There is also research indicating that cigarette smoke not only causes damage in the lungs but also inhibits the reactions of the body to repair itself. An increased chance of heart trouble has also been linked to smoking.

Not all the smoke from a cigarette goes through the cigarette to the smoker; a large percentage of the components of the smoke are exhaled into the air by the smoker. This causes an atmosphere of smoke for other people to breathe. Does this cause illness in nonsmokers exposed to smoke? This is a hotly debated question, and the research that has been done is not conclusive; however, a study in Japan indicated that lung cancer in women increases as the amount of smoking in the home increases. Another study concluded that chronic exposure to tobacco smoke at work is deleterious to the nonsmoker and significantly reduces functioning of small airways. The most frightening results of research are on the effect of secondhand smoke on children. Babies of women who smoke during pregnancy tend to weigh less and develop more slowly than those of nonsmoking mothers. The children of parents who smoke have been found to be more prone to bronchial ailments such as asthma.

In this investigation, we will measure the amounts of solids contained in smoke that are inhaled from various sections of a cigarette as well as what is released into the environment around a smoker.

WASTE AND THE ENVIRONMENT

The damage to the environment that occurs during this investigation is to the air.

▲ *CAUTION*s warn about safety hazards.
*EXTRA*s give helpful hints, additional information, or interesting facts.

Laboratory Equipment
  2 filter flasks
  2 funnels
  1-hole stoppers to fit filter flasks
  2 glass tubing sections to fit stoppers
Common Materials
  filtered cigarettes
  cigarettes:non-filter

eyedropper
ring stand
2 utility clamps
filter paper
accurate scale (to +0.001 g)

matches
little cigar

PROCEDURE
Part I: Determining Solids in Cigarette Smoke

1. Each pair of students should obtain a 250 mL flask, a piece of filter paper, a 1-hole rubber stopper, some glass tubing, a glass funnel, and a piece of rubber tubing. Assemble the equipment as shown and attach the suction flask to a water aspirator.

▲ *CAUTION*
Be careful when putting the glass tubing through the stopper.

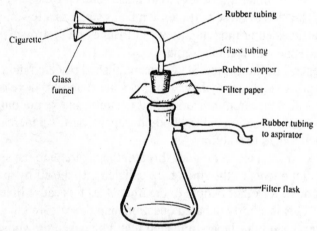

Cigarette

Glass funnel

Rubber tubing

Glass tubing

Rubber stopper

Filter paper

Rubber tubing to aspirator

Filter flask

2. Remove the piece of filter paper from the apparatus and weigh it. **Note: All mass measurements are to 0.001 g.** Record the mass on the report page.

3. Replace the filter paper in the apparatus.

4. Weigh a non-filter tip cigarette.

5. One pair of students should insert the cigarette into the funnel as shown.

6. The other pair of students should simply place the funnel stem into the end of the rubber tubing as shown in the picture below. This will serve to collect room smoke.

7. Clamp the cigarette funnel pointing up and the secondhand smoke funnel about a half an inch above the cigarette funnel. The funnels should be far enough apart to allow air in but close together enough to catch all the smoke.

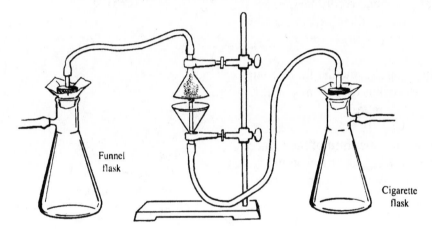

Funnel flask

Cigarette flask

8. Turn on both water aspirators and light the cigarette.

⚠ CAUTION
Don't burn yourself.

9. Adjust the aspirator so that the cigarette burns slowly, over a period of 3 to 5 minutes.

10. When the cigarette has burned down to the last centimeter, turn off the aspirators and put out the cigarette.

11. Remove the filter paper from both suction flasks and weigh each paper.

12. Weigh the unconsumed part of the cigarette. (Do not weigh the ash.)

13. Record all masses on the report sheet.

14. Repeat steps 2 through 13 with a similar cigarette but with puffing simulated by pinching the aspirator hose for 5 seconds and releasing for 2 seconds. Be sure to use new filter papers.

15. Repeat steps 2 through 13 for a filter-tip cigarette using new pieces of filter paper.

16. Repeat steps 2 through 13 for a "little cigar" using new pieces of filter paper.

Part II: Solids from Different Parts of a Cigarette

1. Using a pencil or marking pen, mark off the length of the tobacco part of a filter cigarette in thirds.

EXTRA
Don't poke a hole in the cigarette paper.

2. Weigh two pieces of filter paper and record the masses on the report page. Place one piece of filter paper in each flask.

3. Place the marked filter-tip cigarette into the funnel. Turn on the water. Light the cigarette and as quickly as possible position the funnel over the cigarette.

4. Allow the cigarette to burn until one-third of the tobacco portion has been consumed. Extinguish the cigarette with a drop of water from an eyedropper as you turn off the aspirator.

EXTRA
Use only one drop to extinguish the cigarette. You have to relight it.

5. Weigh both pieces of filter paper and record those masses on the report sheet.

6. Replace the filter paper, turn on the aspirator, and relight the cigarette. Allow the second third of the cigarette to burn. When the second third is consumed, turn off the aspirator and extinguish the cigarette with a drop of water from an eyedropper. Weigh the filter papers and record the masses on the report sheet.

7. Repeat step 6 for the last third of the cigarette.

8. Throw all solid residue in the trash. Be sure the cigarettes and cigar are totally extinguished.

SOLIDS IN SMOKE
PRE-LAB QUESTIONS

Name: _____

Lab Partner: _____

Section: _____   Date: _____

1) How will the measurement of smoke from the burning end of a cigarette differ from the smoke that would be exhaled into a room by a smoker?

2) We will measure all the solids in the smoke that comes from the cigarette. How will this be different than the actual solids a smoker gets from a cigarette?

3) Who do you predict will profit from the use of a filtered cigarette? (the smoker, others in the room, or both the smoker and others in the room)

4) A person who claims they do not inhale and are thereby protected from the risks of smoking are forgetting about something important. What is it?

SOLIDS IN SMOKE
REPORT SHEET

Name: _____

Lab Partner: _____

Section: _____   Date: _____

Second Pair: _____   _____

PART I: SOLIDS IN CIGARETTE SMOKE

	Nonfiltered Cigarette		Puffed Nonfiltered Cigarette		Filtered Cigarette		Little Cigar	
	Cigarette Flask	Funnel Flask	Cigarette Flask	Funnel Flask	Cigarette Flask	Funnel Flask	Cigarette Flask	Funnel Flask
Mass of filter paper (after collecting smoke)	__ g	__ g	__ g	__ g	__ g	__ g	__ g	__ g
Minus (initial)	__ g	__ g	__ g	__ g	__ g	__ g	__ g	__ g
Mass of solids collected	__ g	__ g	__ g	__ g	__ g	__ g	__ g	__ g
Mass of cigarette (initial)	__ g	__ g	__ g	__ g	__ g	__ g	__ g	__ g
Minus (after burning)	__ g	__ g	__ g	__ g	__ g	__ g	__ g	__ g
Mass of cigarette consumed	__ g	__ g	__ g	__ g	__ g	__ g	__ g	__ g

$$\frac{\text{mass of solids}}{\text{mass of cig. consumed}} = \text{\_\_ \_\_} \qquad \text{\_\_ \_\_} \qquad \text{\_\_ \_\_} \qquad \text{\_\_ \_\_}$$

$$\frac{\text{Smoker solids} \times 100\%}{\text{Room solids}} = \text{\_\_ \%} \qquad \text{\_\_ \%} \qquad \text{\_\_ \%} \qquad \text{\_\_ \%}$$

QUESTIONS

1. Which of the cigarettes in Part I produced the most solids in smoker smoke and in room smoke? Can you explain the reason for your results?

2. Which cigarette produced the lowest ratio of smoker smoke to room smoke?

3. What conclusions can you draw from the data about being in a room with a smoker?

267

PART II: EFFECTIVENESS OF A CIGARETTE FILTER

	Smoker Smoke	Room Smoke
1. Mass of initial filter paper	_____ g	_____ g
2. Mass of filter paper after 1/3 cigarette consumed	_____ g	_____ g
3. Mass of filter paper after 2/3 cigarette consumed	_____ g	_____ g
4. Mass of filter paper after last third consumed	_____ g	_____ g
5. Mass of solids from first third (2 − 1)	_____ g	_____ g
6. Mass of solids from second third (3 − 1)	_____ g	_____ g
7. Mass of solids from last third (4 −1)	_____ g	_____ g
8. Ratio for first third Smoker smoke/room smoke	_____	
9. Ratio for second third Smoker smoke/room smoke	_____	
10. Ratio for last third Smoker smoke/room smoke	_____	

EXTRA
Units are not necessary on these numbers, since they are ratios.

QUESTIONS

1. Which portion of the cigarette in Part II yielded the most solids in smoker smoke? Explain your result.

2. Does the portion of the cigarette being smoked affect the amount of solids in the room smoke? If so, why?

3. Does the filter protect a smoker from solids? Does a filter protect someone nearby?

33 *Clarification of Water*

Is It Clean Enough to Drink?

OBJECTIVES

To understand the process of removing particulate matter from water as one of several steps necessary to make it drinkable.

To recycle used aluminum cans into a usable compound.

Relates to Chapters 12 and 14 of Chemistry for Changing Times, *twelfth ed.*

BACKGROUND

Cleaning water sufficiently to make it *potable*, or drinkable, is a major task for most of the world. As the world's population increases, so does the problem. A primary method of cleaning water is to use a *flocculant*, a compound that produces a precipitate that settles to the bottom, trapping suspended solids as it settles. This compound increases the speed at which suspended particles will settle and traps some particles that would not otherwise settle. One example of this type of compound is aluminum hydroxide, a gel-like solid. The aluminum hydroxide increases the amount of solids that settle out and the speed at which they settle out, but the time period is often days or weeks. In this investigation, we will observe a reaction that will cause the aluminum hydroxide to form more quickly.

If the aluminum hydroxide can be made from aluminum cans, litter is also reduced. Aluminum cans could be dissolved by the reaction with a strong base according to the following equation:

$$2\ Al(s) + 2\ Na^+ + 2\ OH^- + 6\ H_2O \rightarrow 2\ Na^+ + 2\ Al(OH)_4^- + 3\ H_2(g)$$

The aluminum complex can be changed to aluminum sulfate by the addition of sulfuric acid:

$$Al(OH)_4^- + 2\ H_2SO_4 \rightarrow 4\ H_2O + Al^{3+} + 2\ SO_4^{2-}$$

It is not necessary to add enough sulfuric acid to dissolve all of the $Al(OH)_3$ formed with the $Al(OH)_4^-$. The surplus $Al(OH)_3$ is removed in the filtration, but having too much sulfuric acid is a problem. Large fluffy masses of aluminum hydroxide are then precipitated from the solution by the addition of sodium hydrogen carbonate (sodium bicarbonate, a cheap base) by the following reactions:

$$Na^+ + HCO_3^- + H_2O \rightarrow Na^+ + CO_2 + H_2O + OH^-$$

$$Al^{3+} + 3\ OH^- \rightarrow Al(OH)_3(s)$$

The filtrate from the aluminum sulfate solution contains aluminum ions in an acidic solution. Adding sodium hydrogen carbonate until the solution becomes basic produces hydroxide ions, which precipitate the aluminum ions as aluminum hydroxide ($Al(OH)_3$), a gelatinous precipitate that will trap particles suspended in water.

In this investigation, water treated with the aluminum hydroxide is compared with dirty water that settles without the precipitate and with the original dirty water. The settling of the visibly "dirty" water is speeded up by centrifuging to accommodate lab time.

Aluminum hydroxide is not usually made from cans, as it is too expensive. Recycling used aluminum cans into new aluminum cans is more economical.

Another method of cleaning water is to pass it over activated charcoal. This is considered an advanced or tertiary method. Activated charcoal works by adsorbing the heavy molecules in the water. The molecules are caught on the irregular surface of the activated charcoal. After the water is cleaned, the charcoal can be regenerated by heating it to 500 to 1000°C with steam or carbon dioxide.

One of nature's methods of cleaning water is the process of crystallization. When a water solution is cooled to below the freezing point, the water will begin to freeze, excluding foreign material from the ice crystals and leaving a more concentrated solution in the liquid state.

WASTE AND THE ENVIRONMENT

Acidic and basic solutions can damage plumbing unless neutralized or diluted. Aluminum ions are not classified as toxic, but it is better not to put metal ions into the water system.

⚠ *CAUTION*s warn about safety hazards.
*EXTRA*s give helpful hints, additional information, or interesting facts.

Reagents

6 M sodium hydroxide [NaOH]

3 M sulfuric acid [H_2SO_4]

sodium hydrogen carbonate
(sodium bicarbonate) [$NaHCO_3$]

manganese dioxide [1 g MnO_2/200 mL]

pH paper (wide) range

ethyl or isopropyl alcohol
[CH_3CH_2OH or $CH_3CH(OH)CH_3$]

Common Materials

aluminum cans, cut into strips 1 cm × 3 cm

fabric dye

activated charcoal

dry ice, broken or crushed

food coloring

dirty water

Laboratory Equipment

fume hood

centrifuge

spectrophotometer set at 500 nm (optional)

metal cutter

2 Buchner filter systems

filter paper

inorganic waste container

3 small test tubes

ring stand, ring, and wire gauze

laboratory burner

125-mL Erlenmeyer flask

150 mL beaker

evaporating dish

100 mL graduated cylinder

10 mL graduated cylinder

stirring rod

PROCEDURE

1. Prepare a cold bath by placing dry ice (solid CO_2) in a 100- or 150 mL beaker with about 50 mL of isopropyl or ethyl alcohol.

2. Place 1 drop of food color in a test tube. Fill the test tube 2/3 full with water. Place the test tube in the cold bath. At 1-minute intervals, observe the water freezing in the test tube.

3. Cut two 1-cm x 3-cm strips from an aluminum can.

4. Place the strips in an evaporating dish with 15 mL of 6 M sodium hydroxide. Make sure the strips are completely covered.

5. Warm the dish with a Bunsen burner in a fume hood. Do not boil. Stir gently to prevent foaming over.

6. Continue warming and stirring until most of the fizzing has stopped (about 5 min). Cool the solution.

7. Add 20 mL of 3 M sulfuric acid and stir well to dissolve as much remaining aluminum hydroxide as possible.

8. Filter the aluminum sulfate solution by suction filtration using the Buchner funnel suction apparatus as shown in the sketch. Wet the filter paper slightly with distilled water to seat the filter before adding the solution.

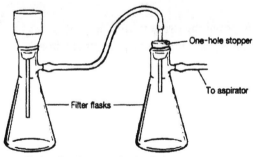

One-hole stopper

To aspirator

Filter flasks

Buchner funnel system with safety flask

9. Add 10 mL of the filtrate to 10 mL of dirty water in a 150 mL beaker. Add solid sodium hydrogen carbonate slowly while mixing until the solution is basic to pH paper (pH > 7). Stir well and pour some of the slurry into a small test tube and mark the tube "1." Note the volume of the aluminum hydroxide formed.

⚠ *CAUTION*
Dry ice can cause frostbite.

⚠ *CAUTION*
The edges are sharp.

⚠ *CAUTION*
NaOH is caustic to the skin.

⚠ *CAUTION*
Don't burn yourself.

⚠ *CAUTION*
Sulfuric acid is corrosive to the skin.

EXTRA
Powered MnO_2 suspended in water makes a good "dirty" water. Hydrogen carbonate is also called bicarbonate.

271

10. Although the solids will settle over a period of time, you can speed the process by centrifuging. Place about the same volume of dirty water in a second test tube marked "2." Balance the centrifuge by placing the test tubes on opposite sides. Centrifuge for 2 minutes.

11. Place the same volume of dirty water in a third test tube marked "3." Compare the three test tubes visually. Compare the three in a spectrophotometer set at 500 nm. Record the absorbance of each.

EXTRA
The lower the absorbance, the clearer the sample.

12. In each of two 125-mL Erlenmeyer flasks, add the same number of drops of fabric dye or food color to about 50 mL of water. Add about 0.5 g of activated charcoal to one flask, stopper, and agitate briskly.

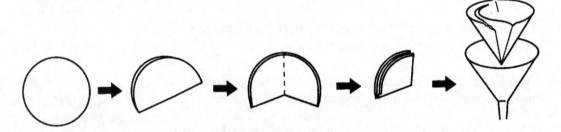

13. Prepare and seat a filter paper in a funnel as shown above. Separate the liquid from the charcoal solid by gravity filtration through filter paper. Compare the filtrate with the dirty water.

14. You may flush the solutions down the drain with plenty of water. Place the solids containing aluminum or manganese in a container for burial at a toxic waste site. Place the charcoal in a container so that it can be reactivated.

CLARIFICATION OF WATER
PRE-LAB QUESTIONS

Name: _____

Lab Partner: _____

Section: _____ Date: _____

1) Why is speed important in the settling of solid particles from water in a municipal water treatment plant?

2) In the initial reaction with the aluminum and sodium hydroxide, what is the source of the hydrogen gas?

3) Aluminum is actually very reactive with water. From the equation representing the first reaction in the process, what substance is allowing the aluminum to react with the water?

4) People who live in wilderness areas or certain rural areas might have to clean their own water, depending on the purity of the source of their water. List the substances and materials that would be necessary for a person to set up their own mini water-treatment plant.

5) In question 3, what final step would you suggest the person take to be certain the water is safe to drink?

CLARIFICATION OF WATER
REPORT SHEET

Name: _____

Lab Partner: _____

Section: _____   Date: _____

I. ALUMINUM HYDROXIDE
Compare the volume of aluminum hydroxide obtained with the original volume of aluminum metal.

II. USE OF ALUMINUM HYDROXIDE
Absorbance of dirty water   _____

Absorbance of centrifuged dirty water   _____

Absorbance of clarified water   _____

III. USE OF CHARCOAL
Absorbance of dirty water   _____

Absorbance of clarified water   _____

IV. QUESTIONS
1. What gas bubbled out of the solution when aluminum reacted with alkali (NaOH)?

2. Is it easy to see aluminum hydroxide suspended in the water? Explain.

3. What gas was formed when sodium hydrogen carbonate was added to the sulfuric acid solution?

4. What happens to the colorful decorations on the aluminum can when you dissolve the can?

5. Suppose the activated charcoal adsorbed a very toxic substance other than a heavy metal. What would you do with the spent charcoal?

6. Is the clarified water you have prepared safe to drink? Why or why not?

7. Why isn't a centrifuge used to clean the city water?

8. Is the fabric dye or food color separated from the water?

34 *Freezing Water*

Which Freezes Faster?

OBJECTIVES

To observe the effect of dissolved gases on the freezing point of tap water.

To demonstrate that boiling removes dissolved gases and that the gases can be reabsorbed during agitation.

Relates to Chapters 14 and 6 of Chemistry for Changing Times, *twelfth ed.*

BACKGROUND

Most of us are familiar with the concept of freezing water. It is known that pure liquid water changes to solid ice at 0.0°C at 1 atm pressure. Adding salt or antifreeze to water lowers the freezing point. Water is denser than ice, so ice floats with about 10% of the ice above water. Since ice is less dense than water, ice must have a larger volume, so ice must swell as it freezes.

There are also some myths about freezing water. You may have heard that an ice company boils water and then freezes the hot water. From that evidence, someone may have said, "That proves hot water freezes faster." Do ice companies boil the water to make it hot, or do they boil it to remove the oxygen and other gases that are dissolved in the water? Perhaps we can answer some of these questions.

Boiling water removes the gases dissolved in it. Pouring the water from one container to another or vigorous stirring allows oxygen and carbon dioxide to re-dissolve in the water. It is a well-known trick of campers to pour boiled water back and forth to remove the flat taste. Everyone has grown accustomed to the taste of water with gases dissolved in it.

Those gases also lower the freezing point of water the same way that antifreeze lowers the freezing point of water in an automobile cooling system. When water is at the freezing point, it is in an equilibrium condition: molecules of solid are becoming liquid, and molecules of liquid are becoming solid. Adding a solute to the liquid diminishes the number of molecules of liquid water at the interface between the liquid and the solid and decreases the number of liquid molecules returning to the solid crystal. Since the same number of solid molecules is going to liquid, the solid size decreases or melts, and a new equilibrium develops at a lower temperature. The addition of a solute has lowered the freezing point of the solvent, in this investigation, water.

In this investigation, it is predicted that the boiled, cooled, and covered water will freeze first (at a higher temperature) because it has no gases dissolved in it. That would explain why the ice company boils its water. The regular tap water and the stirred water should require about the same amount of time because they start at the same temperature and both contain dissolved gases. However, because there are differences in how individual students prepare the test tubes, there are often differences in the order in which the test tubes freeze. The order of freezing depends upon the temperature of the solution, the amount of gas dissolved in the water, and the amount of contact the test tubes have with the cold bath.

Water contains absorbed gases and freezes from the top, allowing fish to survive in frozen ponds and lakes.

WASTE AND THE ENVIRONMENT

The solutions used in this investigation are not toxic.

⚠ *CAUTION*s warn about safety hazards.

*EXTRA*s give helpful hints, additional information, or interesting facts.

Reagents

alcohol (isopropyl or ethyl) [$CH_3CH(OH)CH_3$ or CH_3CH_2OH]

Common Materials

dry-ice pieces   grease pencils

Laboratory Equipment

ring stand, ring, and wire gauze   100 mL beaker
2   250 mL Erlenmeyer flasks   2 watch glass
1000 mL beaker   100 mL graduated cylinder
4 large test tubes   thermometer
stirring rod

PROCEDURE

1. In a 250 mL beaker, boil 100 mL of water for 10 minutes at a medium boil.

2. Pour about half of the boiled water into each of two Erlenmeyer flasks and cover one of the tops with a watch glass.

3. When the two samples have cooled to near room temperature, boil an additional 50 mL of water for 10 minutes at a medium boil. It may be necessary to run tap water over the outside of the two flasks to cool the boiled water to room temperature.

4. Prepare a cold bath in a 1000 mL beaker by adding 100 mL of small pieces of dry ice (solid carbon dioxide) to 300 mL of isopropyl or ethyl alcohol.

5. Prepare four test tubes by numbering them with a grease pencil. Place 10 mL of un-boiled tap water at room temperature in one test tube. Place 10 mL of room-temperature boiled water from the covered flask in the second test tube. Place 10 mL of room-temperature boiled water from the uncovered flask in the third test tube. Stir the water in this test tube rapidly with a stirring rod for 1 minute. Place 10 mL of hot, boiled water in the fourth test tube. Use a thermometer to check that the water samples in the first three test tubes are the same temperature.

6. Record the time. Place the test tubes in the ice bath. The level of the alcohol should be above the level of water in the test tubes. Stir the bath gently. Watch the water for freezing as you move the test tubes back and forth. Record the time when ice first forms across the top of the water in each test tube.

7. Flush the alcohol down the drain with plenty of water. The dry ice can be left to sublime.

⚠ *CAUTION*
Don't burn yourself.

EXTRA
A stopper might stick in the flask due to suction as the water cools.

⚠ *CAUTION*
Dry ice can cause frostbite.

EXTRA
The rapid stirring allows the water to reabsorb gases from the air.

⚠ *CAUTION*
You can freeze your hand to a frosted beaker.

278

FREEZING WATER
PRE-LAB QUESTIONS

Name: _____

Lab Partner: _____

Section: _____   Date: _____

1) Why do fish prefer cooler water?

   More diss gases

2) Use the knowledge from the introductory paragraphs to explain why the hot water pipes in the exterior walls of poorly insulated houses many times freeze before the cold water pipes.

   Less gas

3) In preparing tap water for fish tanks, some people first boil it to remove the chlorine gas and then allow it to sit to cool. What additional step should be taken before expecting fish to be able to survive in the boiled water? Explain.

4) Why is freezing water a concern in cracks of roadways and bridges?

FREEZING WATER
REPORT SHEET

Name: _____

Lab Partner: _____

Section: _____   Date: _____

I. FREEZING WATER

	Initial Time	Time at Freezing	Difference in Minutes
Hot, boiled tap water	_____	_____	_____
Cool, un-boiled tap water	_____	_____	_____
Cool, boiled, covered tap water	_____	_____	_____
Cool, boiled, stirred tap water	_____	_____	_____

II. QUESTIONS

1. In what order did the water samples freeze?

2. What is the effect of the initial temperature on the time of freezing?

    ↑ temp longer

3. What is the effect of dissolved gases on the time of freezing?

    More gases longer

4. Was there a difference in the time required to freeze the tap water and the cool, boiled, stirred tap water? If there was, why?

5. Could fish survive in a pond covered with ice if gases did not dissolve in water?

281

6. Could fish survive in a frozen pond if the water froze from the bottom up?

35 *Freezing Point Depression*
Will Your Radiator Survive the Winter?

OBJECTIVES

To observe the effect of dissolved substances on the freezing point of water.

To determine the density of pure glycerin and several glycerin solutions from mass and volume measurements.

To graph the relationship between density of glycerin solutions and freezing point.

To graph the relationship between percent glycerin and solution density.

Relates to Chapter 14 of Chemistry for Changing Times, *twelfth ed.*

BACKGROUND

The water in an automobile cooling system would freeze when the outside temperature dropped below the freezing point of water without the addition of antifreeze. The damage caused by the frozen water would be a result of the swelling of the water as it froze. Antifreeze lowers the freezing point of the water in the cooling system and thus protects the car from the damage caused by the cold weather to a certain temperature. Many of the commercial antifreezes contain the compound ethylene glycol:

$$HOCH_2CH_2OH$$

The addition of ethylene glycol lowers the freezing point of the water solution in the cooling system. When water is at the freezing point, it is in an equilibrium condition: Molecules of solid are becoming liquid, and molecules of liquid are becoming solid. Adding a solute to the liquid diminishes the number of molecules of liquid water at the interface between the liquid and the solid and decreases the number of liquid molecules going to solid. Since the same number of solid molecules are going to liquid, the amount of solid decreases or melts, and a new equilibrium develops at a lower temperature. The addition of a solute such as ethylene glycol has lowered the freezing point of the solvent—in this investigation, water.

The antifreeze also raises the boiling point of the water and thus helps prevent it from boiling over during the summer. Most commercial antifreezes also contain anticorrosion compounds.

The densities of water and of ethylene glycol are different. As more ethylene glycol is added, the density of the mixture changes. When the antifreeze in an automobile is checked, the measurement made is actually a density measurement.

An ionic compound such as sodium chloride (table salt) would lower the freezing point even more than ethylene glycol does because salt produces more particles per molecule (two ions per molecule instead of one molecule) when it dissolves than ethylene glycol. The ions produced by table salt would also be electrolytes (particles which carry charges in solution), which would cause much more rapid corrosion.

One requirement for an antifreeze is that it mix with water in all proportions. The hydroxyl groups on ethylene glycol make it very soluble in water, and for the same reason water is very soluble in ethylene glycol; that is, they are miscible liquids. The oxygen ends of the water molecules are attracted to the H of the hydroxyl groups on ethylene glycol, and the ethylene glycol becomes surrounded by water molecules.

Because ethylene glycol is very toxic, you will use glycerin (also called glycerol) in this investigation. Glycerin is similar to ethylene glycol in that it lowers the freezing point of water, but it does not lower the freezing point as much as ethylene glycol. The density of glycerin is also very different from that of water. Glycerin has three hydroxyl groups, one on each of the three carbons in the compound. It is used as a softening agent in some foods and in hand lotions. It is also sweet.

$$\text{HOH}_2\text{C}-\overset{\overset{\displaystyle \text{OH}}{|}}{\underset{\underset{\displaystyle \text{H}}{|}}{\text{C}}}-\text{CH}_2\text{OH}$$

Although ethylene glycol is toxic, it is used in commercial antifreezes because it does lower the freezing point more than glycerol. Because ethylene glycol is a toxic compound, it should not be flushed down the drain. Many pets are injured or killed each year by licking antifreeze that has dripped out of a car. It requires only a teaspoon of ethylene glycol to kill a cat. It is reported that ethylene glycol is sweet, as are other compounds with multiple OH groups, such as sorbitol or xylitol. If you see a spot of antifreeze in the garage, wipe it up with a paper towel and throw the towel in the trash.

WASTE AND THE ENVIRONMENT

Glycerin solutions are to be saved and reused. They are not toxic and can be flushed down the drain with plenty of water when disposal is necessary.

⚠ *CAUTION*s warn about safety hazards.
*EXTRA*s give helpful hints, additional information, or interesting facts.

Reagents
  isopropyl or ethyl alcohol [$CH_3CH(OH)CH_3$ or CH_3CH_2OH]
  glycerin [$HOCH_2CH(OH)CH_2OH$]

Common Materials
  dry-ice pieces
  paper towel (or cloth) to wrap graduated cylinder

Laboratory Equipment
  10 mL graduated cylinder
  25 mm x 200 mm test tube
  100- or 150 mL beaker
  dropper
  thermometer
  stirring rod
  centigram balance

PROCEDURE

1. Prepare an air bath system by wrapping strips of paper or cloth around the top of a 10 mL graduated cylinder until it just fits, snugly, into a 25-mm diameter test tube. Measure the mass of the cylinder-in-test tube assembly.

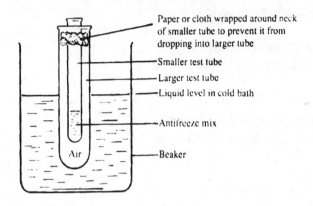

Paper or cloth wrapped around neck of smaller tube to prevent it from dropping into larger tube

Smaller test tube

Larger test tube

Liquid level in cold bath

Antifreeze mix

Air

Beaker

2. Determine the density of glycerin by the following procedure. With the aid of an eyedropper, place exactly 3 mL of glycerin in the cylinder. Weigh and record the mass of the assembly and liquid.

3. Prepare a cold bath by placing dry ice (solid CO_2) in a 100 or 150 mL beaker with about 50 mL of isopropyl or ethyl alcohol.

4. Place air bath assembly containing the 3 mL of glycerin in the cold bath. Make sure the cold alcohol level is above the level of the antifreeze (glycerin) in the inner tube.

5. Wipe a thermometer clean and use it to gently stir the antifreeze until you observe the formation of slush. Read and record the temperature of the antifreeze.

6. Remove the air bath system and melt the antifreeze without diluting it by running tap water over the outside of the tube.

7. Repeat the determination of the freezing point (steps 4 to 5) of your antifreeze solution.

8. Average the two determinations. If the two freezing points are very different (more than 8°), repeat the determination a third time. Average the two closest values.

EXTRA
There are 20 drops to a milliliter.

⚠ CAUTION
Dry ice can cause frostbite.

⚠ CAUTION
A frosted beaker can cause frostbite.

9. Using the data from step 1, calculate the mass of the glycerin. Calculate the density of the mixture by dividing the mass by the volume in milliliters.

10. Determine the densities and freezing points (repeat steps 1 and 4 through 9) of three mixtures: 25% by volume glycerin in water, 50% glycerin in water, and 75% glycerin in water.

11. Graph the data, plotting freezing point (vertical axis) versus % glycerin (horizontal axis). Include 0% glycerin in water at freezing point 0°C.

12. Graph the data, plotting densities (vertical axis) versus the % glycerin (horizontal axis). Include 0% glycerin density as 1.0 g/mL.

13. Return each solution to its proper container so that it can be reused by the next section.

FREEZING POINT DEPRESSION
PRE-LAB QUESTIONS

Name: _____

Lab Partner: _____

Section: _____   Date: _____

1) Ethylene glycol is a covalently bonded molecule and each molecule remains intact as it dissolves in water. What would be the effect on the radiator and engine of an automobile if one used a salt that dissociates into ions (charged particles) to lower the freezing point of the water even more?

2) Could another compound that is not soluble in water be used as antifreeze?

3) Since glycerin and water are completely miscible, enough of the antifreeze can be added until the antifreeze becomes the solvent and the water is the solute. In this case, which freezing point is being lowered technically—that of the water or that of the antifreeze? Explain.

4) From your explanation in question 3, will there be a limit to how low the freezing point of water in a radiator can be pushed? Why?

FREEZING POINT DEPRESSION
REPORT SHEET

Name: _____

Lab Partner: _____

Section: _____   Date: _____

Sample # _____

I. FREEZING POINT	100% Glycerin	75%	50%	25%
First run	_____	_____	_____	_____
Second run	_____	_____	_____	_____
Average	_____	_____	_____	_____

Water with 0% glycerin freezes at 0°C.

Change the average freezing points to the other
common temperature scale
(i.e., °C → °F or °F → °C)

EXTRA
$°F = 9/5°C + 32°$
$°C = 5/9(°F - 32°)$

Average	_____	_____	_____	_____

II. DENSITY	100% Glycerin	75%	50%	25%
Mass of assembly + mixture	_____ g	_____ g	_____ g	_____ g
Minus mass of assembly	−_____ g	_____ g	_____ g	_____ g
Mass of mixture	_____ g	_____ g	_____ g	_____ g
Density = mass in grams/3 mL =	_____ g/mL	_____ g/mL	_____ g/mL	_____ g/mL

Density of water with 0% glycerin is 1.0 g/mL.

III. QUESTIONS

1. From your graphs, determine the density required to protect an automobile's cooling system to −10°C with a glycerin-water mixture.

2. Should you use pure antifreeze in a car radiator? Why or why not?

3. What is the approximate percentage mixture that has the lowest freezing point? What density is that?

4. When a check is made on an automobile radiator's system, the amount of antifreeze is tested by measuring the density of the solution in the radiator. How does the service person know to what temperature a car is protected by measuring density?

36 *Heat Capacity*

The Pan Is Hotter Than the Water

OBJECTIVES

To observe that different materials have unique heat capacities.

To measure the difference in temperature caused by heat capacities of various materials.

Relates to Chapters 14 and 15 of Chemistry for Changing Times, *twelfth ed.*

BACKGROUND

Heat capacity is the amount of heat required to raise the temperature of a substance by 1 C°. *Specific heat* is the heat capacity of 1 g of substance. Water has been used as the standard for heat capacity; thus, its specific heat is 1.00 calorie per gram-degree (4.184 J/g C°). Most of us have some feeling for heat capacity. We know that if a pan of water has been on the stove for thirty seconds, the metal is hot enough to burn our hand but the water is not even lukewarm. Most metals have a specific heat that is about one tenth of the specific heat of water. Metals will become hot ten times faster than water.

The high heat capacity of water is one reason that areas close to large bodies of water will change temperature slowly. Hawaii and other islands will have very consistent temperatures year round. The water of the ocean absorbs the Sun's heat during the day and re-emits the heat into the air when the sun goes down, thereby keeping the temperature from changing drastically.

The equation to calculate the amount of heat involved in a temperature change is:

$$\Delta H = \text{mass} \times \text{specific heat} \times \Delta T$$

where ΔH is the heat gained or lost and ΔT is T (final) − T (initial). The law of conservation of energy also implies that the heat gained by one object is equal to the negative of the heat lost by another object. Thus:

$$\Delta H \text{ gained} = -\Delta H \text{ lost}$$
$$\text{mass} \times \text{specific heat} \times \Delta T = -\text{mass} \times \text{specific heat} \times \Delta T$$

If two equal masses of water at different temperatures are mixed, the temperature of the resulting body of water will be an average of the two temperatures, since the masses and specific heats are equal. However, if the masses are different, the final temperature will be the sum of the products of each mass's temperature times the mass, divided by the total mass. Two different substances placed in contact will both reach a final temperature that lies somewhere between the two initial temperatures. The final temperature will depend on the mass, specific heat, and initial temperature of both objects.

In this investigation, we will combine two masses at different temperatures and observe the final temperature achieved. We will also compare the variations in final temperature as we alter the masses and the substances.

WASTE AND THE ENVIRONMENT

The water/alcohol solutions can be poured down the drain with plenty of water.

⚠ *CAUTION*s warn about safety hazards.
*EXTRA*s give helpful hints, additional information, or interesting facts.

<u>Reagents</u>
  ethanol [CH_3CH_2OH]

<u>Common Materials</u>
  Styrofoam cups
  metal pellets

<u>Laboratory Equipment</u>
  2 thermometers
  2 600 mL beakers
  ring stand, ring, and wire gauze
  laboratory burner
  10 ml graduated cylinder
  small test tubes

PROCEDURE

1. Set up a hot-water bath using a large, half-filled beaker. Adjust the Bunsen burner or hot plate so that the temperature of the water stays almost constant at a temperature between 80°C and 100°C. Set up a cold-water bath by adding ice to water in a half-filled large beaker.

⚠ CAUTION
Don't burn yourself.

2. Place 10 mL of water into each of two test tubes. Place one tube in the hot-water bath until it is the temperature of the bath. Place the other in the cold-water bath until it is the temperature of the bath. Record both temperatures. Empty the contents of both test tubes into a Styrofoam cup. Record the final temperature when it becomes constant.

3. Place 10 mL of water into a test tube. Place the test tube in the cold-water bath. Place 15 mL of water into a test tube. Place the test tube in the hot-water bath. When the temperatures of the water in the test tubes have reached the temperatures of the baths, record both temperatures and then empty the contents of both test tubes into a Styrofoam cup. Record the final temperature when it becomes constant.

4. Place 10.0 mL of ethyl alcohol in the graduated cylinder. Pour the alcohol into a test tube. Place the test tube into the cold-water bath. Place 10 mL of water into a test tube. Place the test tube in the hot-water bath. When the temperatures of the contents of each of the test tubes are the temperature of the baths, record both temperatures. Empty the contents of both test tubes into a Styrofoam cup. Record the final temperature when it becomes constant.

EXTRA
The specific heat of ethyl alcohol is 0.588 cal per gram-degree.

5. Weigh about 10 g of dry metal pellets to the nearest tenth of a gram. Record the mass. Place the pellets into a test tube. Place the test tube in the hot-water bath. Place 10 mL of water into a second test tube. Place the second test tube in the cold-water bath. When the temperatures of the water and the pellets in the test tubes are the temperature of the baths, record both temperatures. Empty the cold water from the test tube into a Styrofoam cup. Carefully and quickly add the pellets to the Styrofoam cup. Record the final temperature when it becomes constant.

EXTRA
The specific heat of the pellets can be looked up in a manual. The instructor will tell you what metal you are using.

6. The pellets can be put back into the pellet container as soon as they are dry. The alcohol–water mixture can be flushed down the drain with plenty of water.

EXTRA
Be careful not to splash as the loss of water will introduce experimental error. Keep the pellets dry while they are being heated.

HEAT CAPACITY
PRE-LAB QUESTIONS

Name: _____

Lab Partner: _____

Section: _____   Date: _____

1) Suppose equal amounts of heat are added to two equal masses of metals at the same initial temperature. Metal A has a specific heat that is half that of Metal B. Which metal will have the highest final temperature? Explain your reasoning.

2) Styrofoam containers are good insulators and transmit very little heat from their contents to the environment. The interior of the Styrofoam maintains the same temperature as its contents. The specific heat of Styrofoam is 0.209 J/g C°, a very small value. Will the heat transferred to the Styrofoam cup present a major error in the laboratory data? Why or why not?

3) Metals are notorious for being excellent conductors of heat. What precautions should be taken to minimize the heat loss from the heated metal pellets in the experiment?

4) If the metal pellets are wet when they are heated and added to the cool water, how will the heat contained in the water on the metal affect the final temperature of the mixture?

HEAT CAPACITY
REPORT SHEET

Name: _____

Lab Partner: _____

Section: _____   Date: _____

I. CALCULATED FINAL TEMPERATURE OF MIXTURES

10 g H_2O and 10 g H_2O $\qquad$ $T_f = \dfrac{T_{hot} + T_{cold}}{2} = \dfrac{{}^\circ C + \quad {}^\circ C}{2} = $ _____ $^\circ C$

10 g H_2O and 15 g H_2O $\qquad$ $T_f = \dfrac{(mass \times T_i)_{cold} + (mass \times T_i)_{hot}}{mass_{cold} + mass_{hot}}$

$$= \dfrac{(10\,g \times \quad {}^\circ C) + (15\,g \times \quad {}^\circ C)}{10\,g + 15\,g} = \underline{\quad\quad} {}^\circ C$$

10 mL ethyl alcohol and 10 g H_2O $\qquad$ T_f

$$= \dfrac{(mass \times sp.\,ht. \times T_i)_{alcohol} + (mass \times sp.\,ht. \times T_i)_{water}}{(mass \times sp.\,ht.)_{alcohol} + (mass \times sp.\,ht.)_{water}}$$

$$= \dfrac{\left(\underline{\quad} g \times \dfrac{0.588\,cal}{g\,{}^\circ C} \times \underline{\quad} {}^\circ C\right) + \left(10\,g \times \dfrac{1.00\,cal}{g\,{}^\circ C} \times \underline{\quad} {}^\circ C\right)}{\left(\underline{\quad} g \times \dfrac{0.588\,cal}{g\,{}^\circ C}\right) + \left(10\,g \times \dfrac{1.00\,cal}{g\,{}^\circ C}\right)}$$

$$= \underline{\quad\quad} {}^\circ C$$

10 g pellets and 10 g H_2O $\qquad$ T_f

$$= \dfrac{(mass \times sp.\,ht. \times T_i)_{pellets} + (mass \times sp.\,ht. \times T_i)_{water}}{(mass \times sp.\,ht.)_{pellets} + (mass \times sp.\,ht.)_{water}}$$

$$= \dfrac{\left(\underline{\quad} g \times \dfrac{\quad cal}{g\,{}^\circ C} \times \underline{\quad} {}^\circ C\right) + \left(10\,g \times \dfrac{1.00\,cal}{g\,{}^\circ C} \times \underline{\quad} {}^\circ C\right)}{\left(\underline{\quad} g \times \dfrac{\quad cal}{g\,{}^\circ C}\right) + \left(10\,g \times \dfrac{1.00\,cal}{g\,{}^\circ C}\right)}$$

$$= \underline{\quad\quad} {}^\circ C$$

II. COMPARISON OF CALCULATED AND MEASURED FINAL TEMPERATURES

	Calculated	Measured
10 g H_2O and 10 g H_2O	_____	_____
10 g H_2O and 15 g H_2O	_____	_____
10 mL ethyl alcohol and 10 g H_2O	_____	_____
10 g pellets and 10 g H_2O	_____	_____

III. QUESTIONS

1. Explain why heat capacity is part of the reason that deserts are hot during the day and cold at night.

2. Considering the specific heats, why is water used instead of alcohol for cooling such things as automobile engines and saw blades?

3. Why is the final temperature of the alcohol and water mixture, and the pellet and water combination, not a numerical average of the two initial temperatures?

4. Which will be hot to the touch when left in the Sun—a piece of metal or a piece of Styrofoam?

5. List any reasons why the calculated temperatures and experimentally measured temperatures are not the same.

37 *Heat of Reaction*

Some Like It Hot

OBJECTIVES

To observe the temperature changes of two reactions, one exothermic and one endothermic.
To calculate the heat of reaction for each of the two reactions using data gathered in the
   investigation.

Relates to Chapter 15 of Chemistry for Changing Times, *twelfth ed.*

BACKGROUND

During a chemical reaction, energy is required to break the bonds between atoms. As new
bonds form, energy is produced. The difference between the energy required and the energy
produced is called the *heat of reaction*. That energy difference is often manifested in a
temperature change of the system.

If, as is the usual case, more energy is produced than is required, the excess energy is given
off as heat. This heat causes the temperature of the chemicals, container, and surroundings to
increase. A reaction of this type is called *exothermic*.

Some reactions will proceed even though more energy is required than is produced. This type
of reaction is called *endothermic* and is characterized by a decrease in the temperature of the
chemicals, container, and surroundings.

Chemists measure the heat of reactions with calorimeters. A *calorimeter* consists of a
reaction vessel enclosed in an insulated space so that little or no heat is transferred to the
surroundings, enabling the experimenter to control energy exchanges. The energy from a
chemical or physical process, as heat, is absorbed by water in the calorimeter and the
calorimeter. The difference in the temperature of the water before and after the reaction is used
to calculate the heat of reaction. Since minimal heat is lost to the environment, the heat of the
reaction is then equal to the heat absorbed by the water and the calorimeter.

The heat absorbed by the water and the calorimeter is calculated by using the equation

$$\Delta H = m_{(H_2O)} C_{P(H_2O)} \Delta t + m_{(calorimeter)} C_{p(calorimeter)} \Delta T,$$

where $m_{(H_2O)}$ is the mass of the water, $C_{P(H_2O)}$ is the specific heat of water, $m_{(calorimeter)}$ is the
mass of the calorimeter, $C_{p(calorimeter)}$ is the specific heat of the calorimeter, and ΔT is the change
in temperature. The product $m_{(H_2O)} C_{P(H_2O)}$ is often called the *heat capacity*. The term *heat
capacity* is also used for the term C_p. The heat of reaction can be determined by calculating the
heat absorbed by each different type of material in the calorimeter and adding the heat absorbed
by the water. In place of the complex calculations, a calibration experiment is usually run. Using
a reaction of known heat of reaction, the calibration experiment determines how much heat the
calorimeter will absorb. The amount of heat absorbed by the calorimeter—its heat capacity—can
then be used to calculate the experimental heat of reaction.

In this investigation, the assumption is made that the calorimeter, nested Styrofoam cups,
thermometer, and test tube do not absorb any heat. That is a fairly good assumption for

Styrofoam, and, fortunately, the test tube and thermometer are not massive and absorb little energy. This assumption simplifies the equation for the heat of a reaction to

$$\Delta H = m_{(H_2O)} \, C_{P(H_2O)} \, \Delta T$$

Another simplifying assumption is that the solutions have the same heat capacity as water. Actually, the solution has the heat capacity of its solute and the water. But for dilute solutions, the mass of water is so much more than the mass of the small amount of solute that the heat capacity of pure water and of the solution are very close to the same value.

In this investigation, both an endothermic and an exothermic reaction will be examined using a simple calorimeter and the simplified equation.

The reaction of sodium hypochlorite and sodium sulfite is exothermic. It is also an oxidation reduction reaction. Sulfite is oxidized to sulfate while hypochlorite is reduced to chloride:

$$ClO^- + SO_3^{2-} \rightarrow Cl^- + SO_4^{2-}$$

The endothermic reaction of ammonium nitrate with water produces a solution of ammonium ions and nitrate ions. The heat of reaction for this process is also called a *heat of solution*.

WASTE AND THE ENVIRONMENT

The hypochlorite/sulfite solution is not toxic. Although nitrate can be detrimental to human health if it is in the water supply, the amount used in this investigation is small.

⚠ *CAUTION*s warn about safety hazards.
*EXTRA*s give helpful hints, additional information, or interesting facts.

Reagents
  1.5 M solution sodium hypochlorite [NaClO]
  1.5 M sodium sulfite solution [Na₂SO₃]
  ammonium nitrate [NH₄NO₃]

Common Materials
  6.4 oz. Styrofoam cup
  8.5 oz. Styrofoam cup

Laboratory Equipment
1   25 x 200 mm test tube
solid stopper to fit test tube
thermometer

PROCEDURE

Part A: Reaction between Sodium Sulfite and Sodium Hypochlorite

1. Prepare a calorimeter from two small Styrofoam cups by cutting two holes in the bottom of the 6.4-oz cup. One hole should be just large enough for the large test tube, and the other hole should be just large enough for a thermometer.

2. Fill the 8.5-oz cup with water to within 5 mm of the top with the large test tube in the cup.

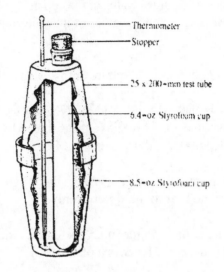

Thermometer
Stopper
25 x 200-mm test tube
6.4-oz Styrofoam cup
8.5-oz Styrofoam cup

3. Place the top cup (6.4-oz) on the calorimeter. It should fit snugly just inside the rim of the bottom (8.5-oz) cup. Place the test tube though the hole and set it against the bottom of the calorimeter. Insert the thermometer into the calorimeter.

4. Place 15 mL of 1.5-M sodium hypochlorite into the test tube. Measure and record the initial temperature.

5. Add 15 mL of 1.5-M sodium sulfite. Stopper the test tube.

6. Gently swirl the cups to mix the two solutions in the test tube. The swirling also transfers the heat to the water evenly.

7. Read and record the temperature at 1-minute intervals until the temperature reaches a maximum and begins to decrease or reaches a plateau.

EXTRA
1.5 M NaClO contains 11 g per 100 mL of water.

EXTRA
1.5 M Na$_2$SO$_3$ contains 21 g per 100 mL of water.

⚠ *CAUTION*
Don't break the thermometer or test tube.

301

8. Remove the test tube and thermometer. Using a graduated cylinder, measure the amount of water in the calorimeter. Also measure the volume of the reaction solution.

Part B: Dissolving Ammonium Nitrate in Water

9. Prepare the calorimeter for another experiment by placing the same amount of room-temperature water in the cup as measured in step **8**. Record the temperature of the water. Rinse and dry the test tube.

10. Weigh out about 10 g of ammonium nitrate onto a watch glass.

11. Place 30 mL of water in the test tube. Add the NH_4NO_3 to the water. Stopper the test tube. Quickly shake the test tube to mix, and place the test tube into the calorimeter.

12. Gently swirl the cups to mix the water. The swirling dissolves the solid and transfers the heat to the water evenly.

13. Record the temperature at one-minute intervals until it reaches a minimum and begins to rise or reaches a plateau.

14. Measure the volume of the test tube contents with a graduated cylinder.

15. You may flush the solutions down the drain with plenty of water.

EXTRA
A 5.25% solution of NaClO is what is sold as bleach.

HEAT OF REACTION
PRE-LAB QUESTIONS

Name: _____

Lab Partner: _____

Section: _____   Date: _____

1) List all the components in the system that will absorb small amounts of heat, besides the water.

2) Considering question 1, will the lab results err on the side of there appearing to be more or less heat exchanged than actual in an exothermic reaction? How about in an endothermic reaction? Explain.

3) Many acids and bases have very high heats of solution. Think of some practical precautions a person should take if they wanted to dilute an acid or base.

4) What will be the effect on the heat measurement in the second reaction if the final temperature is taken before all of the solid dissolves?

HEAT OF REACTION
REPORT SHEET

Name: _____

Lab Partner: _____

Section: _____   Date: _____

I.   SODIUM SULFITE AND SODIUM HYPOCHLORITE

	Temperature		Temperature		Temperature
1. Initial	_____	6	_____	12	_____
1 (minutes)	_____	7	_____	13	_____
2	_____	8	_____	14	_____
3	_____	9	_____	15	_____
4	_____	10	_____		
5	_____	11	_____		

2.   Temperature difference $\Delta T =$ _____ °C $-$ _____ °C $=$ _____ °C
     $\qquad\qquad\qquad\qquad\qquad$ Final $\qquad\quad$ Initial

3.   Volume of water   _____ mL

     Volume of reaction solution   _____ mL

4.   Heat of reaction

$$\Delta H = (\underline{\quad} \text{ mL} + \underline{\quad} \text{ mL}) \frac{1\,g}{mL} \times \frac{1.0\,\text{calorie}}{g\,°C} \times \underline{\quad} °C = \underline{\quad} \text{calorie}$$

$\qquad\quad$ Volume $\qquad$ Volume of $\qquad C_p(H_2O) \qquad\qquad \Delta T$
$\qquad$ of water $\qquad$ reaction
$\qquad\qquad\qquad$ solution

The 1 g/mL is a conversion factor. $C_p(H_2O)$ equals 1.0 calorie per gram degree.

305

II. AMMONIUM NITRATE AND WATER

	Temperature		Temperature		Temperature
1. Initial	_____	6	_____	12	_____
1 (minutes)	_____	7	_____	13	_____
2	_____	8	_____	14	_____
3	_____	9	_____	15	_____
4	_____	10	_____		
5	_____	11	_____		

2. Temperature difference $\Delta T =$ _____ °C − _____ °C = _____ °C

   Final   Initial

3. Volume of water _____ mL

   Volume of reaction solution _____ mL

4. Heat of reaction

$$\Delta H = (\underline{\hspace{1cm}} \text{ mL} + \underline{\hspace{1cm}} \text{ mL}) \frac{1 \text{ g}}{\text{mL}} \times \frac{1.0 \text{ calorie}}{\text{g °C}} \times \underline{\hspace{1cm}} °C = \underline{\hspace{1cm}} \text{calorie}$$

   Volume   Volume of $C_p(H_2O)$ ΔT
   of water   reaction
             solution

III. QUESTIONS

1. How would ignoring the heat absorbed by the Styrofoam cup, the thermometer, and the glass test tube affect the value of the measured heat of reaction?

2. If each compound has its own value of C_p, why do the reaction solutions have the same specific heat as water?

3. Which reaction is exothermic? Explain your reasoning.

38 *Rates of Chemical Reactions*

How Fast Are We Going?

OBJECTIVES

To observe the effects of concentration, temperature, and catalyst on the rate at which a reaction occurs.

To graph the reaction rate as mass versus time.

Relates to Chapter 15 of Chemistry for Changing Times, *twelfth ed.*

BACKGROUND

The rate of a chemical reaction is a measure of how many moles of reactants are used up per unit of time or of how many moles of product are produced per unit of time (for example, in one minute). Three major factors affect how fast a reaction occurs in solution: the concentration of the reactant(s), the temperature, and the presence of a catalyst.

An increase in the concentration of a reactant may cause an increase in the rate of the overall reaction because there are more molecules to react. Increasing the temperature of the reaction causes the reactants to have more energy, and a larger percentage will have an energy greater than the activation energy barrier—that is, more of the molecules will have enough energy to react. The addition of a catalyst allows the reaction to occur at a lower energy level; in other words, it lowers the activation energy barrier. More molecules will have this lower amount of energy and will react, so the reaction goes more quickly or the reaction rate increases.

In this investigation, we will see how each of these variables affects the rate of decomposition of hydrogen peroxide into water and oxygen. The equation is

$$2H_2O_2 \rightarrow 2H_2O + O_2(g)$$

The iodide ion acts as a catalyst. It speeds up the reaction without being used up itself. The more catalyst present, the faster the reaction occurs up to a certain concentration of catalyst. Beyond that concentration the reaction will occur at a rate independent of how much more catalyst is added.

WASTE AND THE ENVIRONMENT

The solutions are all environmentally safe.

⚠ *CAUTION*s warn about safety hazards.

*EXTRA*s give helpful hints, additional information, or interesting facts.

Reagents

0.1 M sodium iodide solution [NaI]

30% hydrogen peroxide [H_2O_2]   3% hydrogen peroxide [H_2O_2]

Laboratory Equipment

timers or watches with second hands   rulers

digital analytical balance ($\pm$ 0.001g)   hot-air dryers

50 mL graduated cylinder

PROCEDURE

Each pair will be assigned a specific amount of 0.1-M sodium iodide (NaI) to be used in the second step. They will later share their data with the class.

1. Select a 50 mL or a shortened, dry 100 mL graduated cylinder. Place exactly 25 mL of fresh 3% hydrogen peroxide (H_2O_2) in the cylinder. Place the cylinder on the balance and start the timer. Be certain to close all cabinet doors on the balance to prevent the air currents from affecting the weighing. Every minute for 15 minutes, record the mass to the nearest 0.001 g in the table on the report sheet. Is there evidence of any decomposition? Do bubbles form?

2. Repeat step 1, except add the assigned amount of 0.1 M sodium iodide (NaI) to the hydrogen peroxide in the cylinder. What does the solution look like?

3. Repeat step 1 using 25 ml of fresh 30% hydrogen peroxide in the cylinder. What does the solution look like?

4. Graph your data, plotting mass (vertical axis) versus time (horizontal axis) on graph paper. From the data at 8 minutes and 12 minutes, calculate the weight loss per minute (a measure of the rate of the reaction). Inform your instructor of the value, and record values from the other groups. Also calculate the rate per minute per milliliter of NaI added and write it on the blackboard with your initials. The instructor will find an average value for the class and calculate a standard deviation.

5. Next, select a large, dry test tube and place 25 mL of 3% H_2O_2 in it. Add 2 mL of 0.1-M NaI and heat to about 35°C . Compare how fast bubbles are evolved at 35°C with how fast bubbles are evolved at 25°C (room temperature).

6. Flush the solutions down the drain with plenty of water.

⚠ *CAUTION*
H_2O_2 is an oxidizing agent.

⚠ *CAUTION*
Hydrogen peroxide is hazardous to skin and clothing.

EXTRA
Put the cylinder onto the balance quickly after adding the NaI.

⚠ *CAUTION*
30% hydrogen peroxide is corrosive. Immediately wash off any spills.

⚠ *CAUTION*
Don't heat too fast. Don't point the test tube at anyone.

RATES OF CHEMICAL REACTIONS
PRE-LAB QUESTIONS

Name: _____

Lab Partner: _____

Section: _____   Date: _____

1) Reactions are often influenced by temperature. How do we slow reactions involving food in our homes?

2) All reactions require an input of energy, the activation energy, to begin. The added energy can come from different sources—friction, heat, light, an electrical spark, or a current. In some reactions, once some of the molecules have enough energy to get over the energy barrier, the reaction proceeds at a very fast rate. What method do we use to give the molecules on the tip of a match enough energy to get over the energy barrier?

3) Most of the biochemical reactions that are necessary for life occur in solution. Why does our body temperature rise when we are ill?

4) What is the name we give to a catalyst in many biochemical and physiological processes?

5) How does the concentration of oxygen in your blood affect the rates of respiratory reactions within the cells?

RATES OF CHEMICAL REACTIONS
REPORT SHEET

Name: _____

Lab Partner: _____

Section: _____ Date: _____

I. Observations of H_2O_2 before adding NaI

II. Volume of 0.1 M NaI added _____ mL

Time (minutes)	3% H_2O_2 Mass (g)	3% H_2O_2 + NaI Mass (g)	30% H_2O_2 Mass (g)
0			
1.0			
2.0			
3.0			
4.0			
5.0			
6.0			
7.0			
8.0			
9.0			
10.0			
11.0			
12.0			
13.0			
14.0			
15.0			

        3% H_2O_2 3% H_2O_2 + NaI   30% H_2O_2

III. Rate $= \dfrac{W_8 - W_{12}}{t_{12} - t_8} =$ _____ g/min   _____ g/min   _____ g/min

IV. CLASS VALUES

        2 mL of NaI   _____ g/min

        4 mL of NaI   _____ g/min

        6 mL of NaI   _____ g/min

        8 mL of NaI   _____ g/min

V. What is the rate per minute per milliliter of NaI?

VI. QUESTIONS

1. Did heating change the reaction rate?

2. What would be the effect of cooling H_2O_2 in a refrigerator?

3. What effect did increasing the concentration of the catalyst (NaI) have on the reaction rate?

4. What was the decomposition rate before the addition of the catalyst?

5. Were these chemicals environmentally safe?

6. What is an appropriate way to store H_2O_2?

INVESTIGATION

39 *Canned Fuel and Evaporation*

Up, Up, and Away

OBJECTIVES

To observe several factors affecting the rate of evaporation of a volatile liquid.

To measure heat loss from the surroundings as a liquid evaporates.

To produce a fuel composed of a volatile liquid in a gel and to use this gel as a heat source for another section of the investigation.

Relates to Chapters 6, 14, and 15 of Chemistry for Changing Times, *twelfth ed.*

BACKGROUND

Humans are becoming more aware of the depletion of fossil fuels. Research programs are currently searching for alternative sources of energy, such as nuclear, solar, tidal, and biomass.

Other programs are looking into stretching fossil fuels by conservation or by the use of additives. For example, alcohol is added to gasoline to make "gasohol." E-10 is a form of gasohol that is 10% ethanol and 90% gasoline and E-85 is 85% ethanol. There are also programs to promote the use of alcohol instead of gasoline in automobiles and planes. Ethanol is being investigated for use in fuel-cell technology.

Alcohol can be produced synthetically or by fermentation. The alcohol produced from fermentation of plant material is part of the *biomass* energy source. Biomass is the collective mass of biological organisms, both dead and alive, excluding those that have been converted into substances such as coal or oil. Biomass is stored solar energy. Beginning with photosynthesis, radiant energy is stored as energy-rich plant material which then becomes the starting point of the food chain where the energy is passed from organism to organism. Biomass can be burned as a fuel as with peat, manure, wood, grasses, and other plant chaff, or it can be converted into other forms of fuel such as alcohol. Ethanol produced from the sugars and starches of maize and sugarcane is called "bioethanol."

Synthesized alcohol is not an energy source; it requires energy to produce the alcohol. Synthesized alcohol is a fuel made from one of the fossil fuels. It is currently more energy/cost efficient to synthesize ethanol from fossil fuels than it is to ferment it from plant material. The advantage of fermentation is that plant material is a sustainable resource.

In this investigation, you are going to make a semisolid or gel that will burn. Actually, it is the alcohol that is held in the gel that burns. The gel is similar to sterno. If the gel is left uncovered, the alcohol will evaporate. Many good fuels are *volatile* liquids. That is to say that they vaporize easily. As a vapor, the effective surface area of a substance is extremely large and thus increases the reaction rate—a necessary attribute for a good fuel to be used in engines where rapid combustion is required.

Evaporation is a phenomenon that many of us have observed. At least we have observed the reduction in the volume of a liquid over time. Liquid is lost because some of the molecules escape into the gas phase. At the same time, some gas molecules travel into the liquid and return to the liquid phase. In a closed container, the rate of gas molecules returning to the liquid phase equals the rate of liquid molecules escaping into the gas phase. This is a system at *equilibrium*— a condition where two opposing processes occur at the same rate. In this situation, the gas pressure above the liquid is called the *equilibrium vapor pressure*. In an open container, some gas molecules may escape completely from the area above the liquid, reducing the number of molecules that return to the liquid. Thus, when a volatile liquid is left in an open container, the liquid will continue to evaporate until the liquid in the container is gone. If the gas molecules are removed more quickly than by their own natural motion, fewer are allowed to return to the liquid, and the liquid evaporates more rapidly. Fanning or using an aspirator will remove molecules from the area above the liquid, so the liquid should evaporate more rapidly.

Heat from the surroundings supplies the energy for the liquid molecules to escape to the gas phase. Increasing the temperature will increase the rate at which liquid molecules leave and thus increase the rate of liquid loss.

During normal evaporation, the more energetic or warmer molecules escape into the gas phase, leaving the cooler molecules in the liquid phase. The environment around the solution is cooled because of the heat absorbed to affect the evaporation. When a cup of water is left out to evaporate, the liquid in the cup will cool as evaporation occurs. The water-cooler type of air conditioner uses this concept. The water in the outer pads evaporates, cooling the air drawn through the pads. The cooler air is blown into the room. The water cooler does not work as well in humid climates, because the water evaporates more slowly.

One method of measuring the humidity in air is to compare the temperature readings on two thermometers: one with wet cotton covering the bulb and the other unaltered. The bigger the difference in temperatures after they have been spun around in the air, the smaller the percent humidity. This is because the presence of water in the air restricts how much and how fast more water will evaporate.

A barrier such as a lid can prevent evaporation. A chemical that will not evaporate and will float on the surface of a liquid that does evaporate can also act as a barrier to evaporation.

WASTE AND THE ENVIRONMENT

The gel does not contain materials that are hazardous to the environment, but combinations of solids and liquids are not supposed to be buried. However, if the alcohol is allowed to evaporate, the residue will not be a hazard to the environment.

⚠ *CAUTION*s warn about safety hazards.
*EXTRA*s give helpful hints, additional information, or interesting facts.

Reagents
calcium acetate [$Ca(C_2H_3O_2)_2$]   phenolphthalein
6 M sodium hydroxide [NaOH]   acetone
ethanol [CH_3CH_2OH](denatured)
Common Materials
mineral oil   matches
cotton balls   rubber bands
Laboratory Equipment
10 mL graduated cylinder   50 mL beaker
100 mL graduated cylinder   150 mL beaker
watch glass   funnel with vacuum tubing

PROCEDURE
Canned Fuel

1. Prepare 8 mL of saturated calcium acetate solution by adding 3.2 g of calcium acetate, $[Ca(C_2H_3O_2)_2]$, to 8 mL of water in a 50 mL beaker. Stir until as much as possible dissolves. Add 2 drops of 6 M sodium hydroxide (NaOH).

2. Measure out 60 mL of ethanol into a 150 mL beaker. Add 3 to 4 drops of phenolphthalein.

3. Pour the alcohol into the calcium acetate. If the solutions do not gel immediately, pour the solution back into the alcohol beaker. After the solution gels, pour any remaining liquid into a waste beaker.

4. Ignite the gel. The flame will be difficult to see. To put out the flame, put a watch glass or a textbook over the beaker. The flame will go out before the cover will burn. When the gel is not burning, it should be covered to prevent the alcohol from evaporating.

Evaporation

5. Place 0.5 mL of ethyl alcohol on a watch glass in an area of the lab where there are no drafts. Record the time. Watch the alcohol and record the time to complete evaporation.

6. Place 0.5 mL of ethyl alcohol on a watch glass. Add 1 drop of mineral oil to the surface of the alcohol. Periodically check to see if the alcohol has evaporated.

7. Place 0.5 mL of ethyl alcohol on a watch glass and record the length of time to complete evaporation while fanning the watch glass with a large notepad or a fan.

8. Place 0.5 mL of ethyl alcohol on a watch glass. Using a clear glass funnel on the end of an aspiration hose, cover the watch glass and record the time to complete evaporation while the aspirator reduces the air pressure above the liquid. Touch the funnel to the watch glass on one side only; there must be a small opening for air to be drawn through.

9. Place 0.5 mL of ethyl alcohol on a watch glass on a ring stand about 8 cm above the alcohol gel. Light the gel and record the time to complete evaporation of the alcohol in the watch glass.

EXTRA
Heating this solution will cause less to dissolve.

⚠ *CAUTION*
NaOH is caustic.

EXTRA
The phenolphthalein is added to give the gel a pink color.

⚠ *CAUTION*
Don't burn yourself.

⚠ *CAUTION*
Alcohol is flammable.

EXTRA
The alcohol may never evaporate.

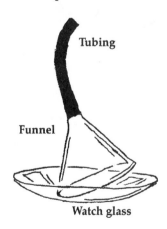

Tubing

Funnel

Watch glass

10. Wrap a cotton ball around the bulb end of a thermometer. Secure the cotton ball in place with a rubber band. Dip the cotton ball into a test tube of ethyl alcohol. Read and record the time and the temperature. Wave the thermometer gently in the air until the alcohol evaporates. Read and record the time and the temperature.

11. Repeat step 10 using acetone instead of ethyl alcohol.

12. Place any leftover gel in an open container in the hood. When the alcohol is evaporated, you may place the residue in the trash.

⚠ CAUTION
Acetone is flammable.

316

CANNED FUEL AND EVAPORATION
PRE-LAB QUESTIONS

Name: _____

Lab Partner: _____

Section: _____   Date: _____

1) How would the rate at which the ethyl alcohol burns be different if it were not held in the gel?

2) If a canned fuel is kept tightly sealed, it will last almost indefinitely. Does that mean that no alcohol is evaporating from the surface of the gel? Explain your answer.

3) The "heat index" is a measurement used by weather forecasters to give us an indication of how the combined effects of temperature and humidity affect our bodies. Why does it feel hotter in the summer when the humidity is high?

4) How will our measurement of the energy absorbed by the ethyl alcohol as it evaporates be affected by atmospheric humidity? Explain your reasoning.

CANNED FUEL AND EVAPORATION
REPORT SHEET

Name: _____

Lab Partner: _____

Section: _____   Date: _____

I. INSTRUCTOR'S INITIALS FOR ALCOHOL GEL _____

II. EVAPORATION

Sample	Start Time	Finish Time	Difference
Alcohol	_____	_____	_____
Oil-covered alcohol	_____	_____	_____
Fanned alcohol	_____	_____	_____
Alcohol in a vacuum	_____	_____	_____
Heated alcohol	_____	_____	_____

III. COMPARISON OF RATES OF EVAPORATION

Sample	Time Start	Finish	Difference	Temperature Start	Finish	Difference
Alcohol	_____	_____	_____	_____	_____	_____
Acetone	_____	_____	_____	_____	_____	_____

IV. QUESTIONS
   1. Why was the NaOH added to the canned fuel? Why was the phenolphthalein added?

   2. Would this gel work as a fuel source in a car? Why or why not?

3. Besides the ethyl alcohol sample with the oil barrier, which sample of alcohol evaporated the slowest?

4. Why did each of the others evaporate more quickly?

5. Why do fanning and the aspirator have a similar effect on the evaporation rate of the ethyl alcohol?

6. Did the material that evaporated the fastest cause the greatest difference in temperature?

7. If acetone could be held in a gel, would it burn at the same rate as ethyl alcohol? Explain the basis for your answer.

40 *Batteries*

Keep Those Electrons Moving

OBJECTIVES

To produce two voltaic cells, one a conventional cell and one from a citrus fruit.

To gain an understanding of the function of a salt bridge in the movement of charged particles from one solution to another in an electrolytic cell.

To measure the potential across the electrodes of each cell.

Relates to Chapters 8 and 15 of Chemistry for Changing Times, *twelfth ed.*

BACKGROUND

Batteries or voltaic cells are arrangements of materials that force electrons through an external wire, forming a circuit. Light bulbs, motors, or radios can be attached to the circuit so that the flow of electrons will cause the bulb to light, the motor to run, or the radio to play.

The flow of electrons is caused by a chemical reaction, called an oxidation-reduction reaction. Each metal has its own unique attraction for electrons. Some metals will pull electrons from other metals. This process reduces the charge on a metallic ion and the tendency to be reduced is called a *reduction potential*. A listing of reactions between metals and their ions in order of increasing potential is called a table of "Standard Reduction Potentials." The relative potentials can be used to design and construct batteries of varying voltages. If two metals are arranged properly, electrons will be pulled through an external wire. The wire is attached at one end to the metal that is pulling the electrons. This metal has the higher reduction poential. When the electrons arrive at this metal electrode, called the *cathode*, the electrons react with *cations* (positively charged ions) in the solution touching the metal electrode. The cations become metal atoms and attach to or "plate" onto the electrode.

The other end of the external wire is attached to the other metal electrode or *anode*. The electrons are generated as metal atoms lose their electrons. The metal atoms become positive ions, cations, in the solution touching the anode. The two solutions have to be connected electrically—that is, ions have to move from one solution to the other so the charge will stay balanced. More specifically, one ion has to move for each electron that travels through the wire. One device that allows the ions to move from one solution to the other is a *salt bridge*. The solution also contains *electrolytes* (or ions) to carry the charge. An acid often is used as an electrolyte in the solutions around the electrodes.

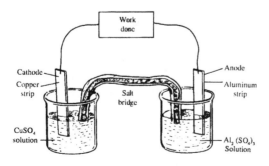

In this investigation, you will build several batteries. In one battery, you will use copper metal as the cathode. The reaction at the cathode is called a *reduction reaction* because the charge on the ion is reduced:

$$Cu^{2+} + 2\ e^- \rightarrow Cu.$$

That is, copper(II) ions gain two electrons to produce copper metal. The electrons come through the wire, and the ions through the solution. At the anode, an *oxidation reaction* occurs where the charge on an atom rises from zero to a positive value:

$$Al \rightarrow Al^{3+} + 3\ e^-.$$

The aluminum atoms lose three electrons and become aluminum ions dissolved in the solution. As copper(II) ions are removed from one solution and aluminum ions are added to the other solution, ions must travel through the salt bridge to keep the solution charges balanced.

In a citrus battery, the pulp and acidic juice of the lemon act as the salt bridge. The anode reaction will be the same as in the previous battery, but the cathode reaction must be different because there are no copper(II) ions. The citric acid supplies hydrogen ions, H^+, to be reduced to hydrogen gas.

$$2H^+ + 2\ e^- \rightarrow H_2\ (g)$$

A voltaic cell can be made using coins for the electrodes. Paper soaked in sodium chloride acts as the salt bridge. The voltmeter is part of the external wire. Since different coins are composed of different metals, there will be a pull of electrons by one coin from the other. If two pairs of coins are stacked in an alternating pattern, it would be as if two batteries were hooked together and were roughly twice the voltage. Our 12-volt car batteries are six lead cells each producing about 2.2 volts.

WASTE AND THE ENVIRONMENT

Although copper and aluminum compounds are not classified as toxic, it is better not to put any metal ions into the water system.

⚠ *CAUTION*s warn about safety hazards.
*EXTRA*s give helpful hints, additional information, or interesting facts.

Reagents
  1 M copper sulfate [$CuSO_4$]   potassium sulfate [K_2SO_4]
  concentrated sulfuric acid [H_2SO_4]   copper metal strips
  1 M aluminum sulfate [$Al_2(SO_4)_3$]   small-gauge copper wire
  sodium chloride [NaCl]
Common Materials
  aluminum foil   cotton balls
  oranges, lemons, or grapefruits   paper clips
  steel wool, fine   clean, shiny coins (penny, nickel, dime)
  paper towels
Laboratory Equipment
  voltmeter   flexible clear plastic tubing (12 cm)
  3 30 mL beakers   10 mL graduated cylinder
  100 mL graduated cylinder

PROCEDURE
Part A: Salt Bridge Battery

1. Into a 30 mL beaker, pour 25 mL of 1 M copper(II) sulfate. Add 1 mL of concentrated sulfuric acid. This is the cathode compartment.

2. Into a second 30 mL beaker, pour 25 mL of 1 M aluminum sulfate. Add 1 mL of concentrated sulfuric acid. This is the anode compartment.

3. To prepare the salt bridge, make a solution of 3 g of potassium sulfate (K_2SO_4) in 30 mL of water. Pour the solution into a 9- to 12-cm.-long piece of clear flexible tubing. Plug the ends of the tubing with cotton. Make sure that the solution is continuous in the tubing with no breaks.

4. Place one end of the salt bridge in the anode compartment and one end in the cathode compartment.

5. Polish the copper metal electrode with steel wool. Place the copper electrode in the cathode compartment.

6. Polish the aluminum metal electrode with steel wool. Place an aluminum electrode in the anode compartment.

7. Call the instructor to measure the current and voltage.

8. Record your observations about the color of the electrodes and the color of the copper solution. Use a wire to connect the two electrodes. Note whether an observable reaction occurs at the electrodes.

9. Set the battery aside to continue discharging while making the citrus cell.

Part B: Citrus Cell

10. Cut a lemon or grapefruit in half across the segments. Push a polished or cleaned aluminum electrode at least 2 cm into the pulp of the fruit.

11. Push a polished copper electrode at least 2 cm into the pulp of the fruit as far from the other electrode as possible but in the same segment of fruit (on a longitudinal line from navel to stem).

⚠ CAUTION
Sulfuric acid is corrosive.

EXTRA
The electrodes can be held in place with a paper clip.

EXTRA
Zinc and zinc sulfate or magnesium and magnesium sulfate can replace aluminum.

12. Take your citrus cell to the instructor to have the current and voltage measured.

13. Reposition the copper electrode to another segment of the fruit to show that the segment dividers act as insulators.

14. Observe the electrodes and the color of the solutions around each electrode in the salt bridge battery.

15. With an eyedropper or pipet, remove one half of each
. solution into a separate beaker. Add distilled water to the battery solutions to the original volume. Ask the instructor to measure the voltage and current.

16. Evaporate the solutions to dryness and place the dry salts in containers to be buried.

Part C: Coin Batteries

17. Make a 1 M solution of NaCl by adding 0.4 g of NaCl to 10 mL of water in a small beaker. Cut 4 circles of paper towel slightly larger than the coins. Soak the circles in the NaCl solution.

EXTRA
The coins should be clean.

18. Make a coin battery by placing a coin, a circle of NaCl-soaked paper towel, and a different type of coin in a stack. The coins must not touch. Ask the instructor to measure the voltage.

EXTRA
Attach the clips of the voltmeter so that the voltages are positive.

19. Make a different coin battery by using a different combination of coins. Ask the instructor to measure the voltage.

20. Make a double coin battery by placing 2 identical pairs of coins in a stack in the order 1, 2, 1, 2 with soaked paper towel circles between each of the 4 coins. Ask the instructor to measure the voltage of the battery.

BATTERIES
PRE-LAB QUESTIONS

Name: _____

Lab Partner: _____

Section: _____   Date: _____

1) Why must two different metals be used for the two electrodes in a battery?

2) Would you expect the voltage in a voltaic cell to increase or decrease the further apart the two metals appear on a table of Standard Reduction Potentials? Explain your reasoning.

3) What will happen if the solution inside the salt bridge develops a gap? Why?

4) What acid acts as an electrolyte in a citrus battery?

5) If the two coins touch without the wet paper towel between them, what voltage will you expect to read from the external wire connecting them? Why?

BATTERIES
REPORT SHEET

Name: _____

Lab Partner: _____

Section: _____   Date: _____

I. SALT BRIDGE BATTERY

Voltage _____ V   Current _____ mA

HALF CONCENTRATION SALT BRIDGE BATTERY

Voltage _____ V   Current _____ mA

II. CITRUS CELL

Voltage _____ V   Current _____ mA

III. COIN BATTERY

Coins _____ and _____

Voltage _____ V   Current _____ mA

Coins _____ and _____

Voltage _____ V   Current _____ mA

Double battery

Coins _____ and _____

Voltage _____ V   Current _____ mA

IV. QUESTIONS

1. Did the intensity of the color of the copper solution change with time?

2. Does the copper electrode, cathode, have a coating or plating on it?

3. If the copper solution touched a metal paper clip, what happened?

4. Why do the different combinations of coins produce a different voltage?

5. Is the double battery twice the voltage of the single battery? Why?

6. What happens if we use two coins of the same metal to make a coin battery?

7. Did the change of concentration in the salt bridge battery result in a change of the voltage?

8. How are the results of halving the concentration in the salt bridge battery and changing the number of coins in the coin battery related to each other?

INVESTIGATION

41 *Test for Vitamin C*

Had Your Ascorbic Acid Today?

OBJECTIVE

To discover various factors that alter vitamin C content in a solution.
To use the technique of titration to determine the amount of vitamin C in a solution.

Relates to Chapters 16, 17, and 19 of Chemistry for Changing Times, *twelfth ed.*

BACKGROUND

Ascorbic acid, or antiscorbutin, is more commonly known as vitamin C. Sometimes called the antiscurvy vitamin, a deficiency of vitamin C leads to scurvy, a disease characterized by loosening of the teeth and small hemorrhages throughout the body. The body uses vitamin C in several ways, including the repair of injured tissue. The body also absorbs extra vitamin C during periods of infection. Excess vitamin C is excreted from the body. Because vitamin C is not stored in the body, good health requires an adequate supply each day.

A balanced diet supplies an adequate amount of vitamin C because it exists in many foods. The most common source is citrus fruits, but vitamin C is also present in acerola berries, rose hips, and white potatoes.

Vitamin C content	mg Ascorbic Acid per 100 g of juice	mg/6 fl oz
Orange juice		
fresh	37 to 61 (type and season variations)	65 to 108
reconstituted, frozen	45	79
canned	40	71
Pineapple juice, canned	9	16
Grapefruit juice, canned, unsweetened	34	60
Pineapple-orange drink (40% juice)	16	28
Tomato juice, canned	16	28

The minimum daily requirement (MDA) for a man, woman, or child is 30 mg of ascorbic acid per day. The recommended daily allowance (RDA) is larger, at 60–70 mg per day.
The molecular formula of ascorbic acid is

The molecular weight is 176.1. Ascorbic acid is very soluble in water, slightly soluble in alcohol, and insoluble in nonpolar solvents like benzene and chloroform. Since ascorbic acid is a strong reducing agent, the test for its presence is usually a reaction with an oxidizing agent like indophenol

329

or iodine. In the reaction, carbonyl (C=O) groups are produced from the OH groups bonded to the ring structure. The double bond of the carbonyl group is changed to a single bond and the hydrogen is lost. Overall, vitamin C loses two hydrogens and two electrons.

Iodine is not soluble in water; therefore, an iodine solution is made by dissolving the iodine molecule (I_2) in a solution of iodide ion (I^-) to make the triiodide ion (I_3^-). When triiodide reacts with vitamin C, it accepts two electrons to become three iodide ions (I^-).

Vitamin C is easily destroyed by heat or contact with copper metal. Oxygen will also destroy vitamin C by an oxidation-reduction reaction similar to the reaction with indophenol or iodine.

This investigation uses the principles of *titration*. One solution is slowly added to another until both solutions have reacted completely. At this point, an *indicator* changes color to denote the *endpoint* of the titration. Burets are often used in titrations, but in this investigation squeeze bottles are used because it is easier to learn to use a squeeze bottle.

The first step of a titration is a *standardization*, in which the amount of iodine needed to react with 1 mg of vitamin C is determined. A known amount of vitamin C is used to make the vitamin C solution. That solution is then titrated with the iodine solution using starch as an indicator.

The next step is to determine the amount of vitamin C in a sample. The sample is titrated with iodine to the endpoint. The amount of vitamin C is calculated from the amount of iodine added.

Starch serves as an indicator in iodine reactions because iodine molecules fit into the starch molecules in such a way that they cause a blue color. It appears that the iodine molecule is trapped inside a coil of the starch molecule. As long as the iodine reacts with vitamin C, there is no iodine to react with the starch. As soon as all the vitamin C has reacted, the next drop of iodine reacts with the starch indicator to produce a blue color. The appearance of the blue color signals the endpoint of the titration.

WASTE AND THE ENVIRONMENT

The vitamin C and starch solutions are not toxic. The nitric acid can cause problems with the plumbing if the acid is not neutralized or diluted.

⚠ *CAUTION*s warn about safety hazards.
*EXTRA*s give helpful hints, additional information, or interesting facts.

Reagents

6 M acetic acid [CH_3COOH]	iodine solution
1% starch solution	6 M nitric acid [HNO_3]
copper metal strips or pieces	distilled water

Common Materials

vitamin C tablets, 250 mg	apple juice or orange juice with vitamin C
citrus juices or powdered drink mixes	

Laboratory Equipment

filter paper	ring stand, ring, and wire gauze
10 mL volumetric pipet	laboratory burner
250 mL volumetric flask	centigram balance
125-mL Erlenmeyer flask	large test tube
100 mL graduated cylinder	small squeeze bottle
250 mL beaker	
100 mL beaker	

PROCEDURE

1. Place about 150 mL of water into a 250 mL beaker. On a ring stand with a ring and wire gauze, heat the water to boiling over a laboratory burner.

⚠ *CAUTION*
Don't burn yourself.

2. Dissolve a 250 mg vitamin C tablet in distilled water in a 250 mL volumetric flask. Fill the flask exactly to 250 mL with distilled water using a dropper for the last several drops.

3. Place 25 mL of the juice to be tested in a large test tube and place it in the boiling water bath for 20 minutes. Set it aside to cool.

4. Place 25 mL of the original vitamin C solution from step 1 in a 100 mL beaker. Clean the copper pellets (or strips) by dipping them quickly in 6 M nitric acid and then in water. Add the pellets or strips of copper metal to the beaker.

⚠ *CAUTION*
Nitric acid is corrosive.

5. Put 25 mL of distilled water into a 125-mL Erlenmeyer flask. Add 2 mL of 6 M acetic acid. With a volumetric pipet, add 10.00 mL of the vitamin C sample prepared in step 1. Add 3 or 4 mL of 1% starch solution to the flask. (Stir the starch solution very well before dispensing because the starch may settle to the bottom of the solution.)

⚠ *CAUTION*
Use a safety bulb with the pipet.

6. Put about 50 mL of the iodine solution in a squeeze bottle. Weigh the bottle (initial weight). Titrate the solution with iodine to the blue (or purple or green) endpoint by adding iodine from the squeeze bottle until the blue endpoint persists for at least 20 to 30 seconds. Reweigh the bottle (final weight). The report sheet will help you to calculate the milligram vitamin C per milliliter iodine.

EXTRA
At the start of the titration, large squirts may be added. You will see streaks of blue that will fade. Close to the endpoint, add single drops.

7. Record the type of juice sample that you will test on the report sheet. The sample can be orange juice or drink, lemon or lime drink powder, or apple juice. Because of their color, tomato juice, grape juice, and other powdered drink flavors cannot be

used in this determination. The color interferes with recognition of the endpoint of the titration.

8. If the sample contains pulp, it should be filtered. Put 25 mL of distilled water into a 125-mL Erlenmeyer flask. Add 2 mL of 6 M acetic acid. With a volumetric pipet, add 10.00 mL of the juice sample. Stir the starch solution very well and add 3 or 4 mL of 1% starch solution to the flask.

9. Weigh the filled squeeze bottle (initial weight). Titrate the solution from step 6 with iodine to the blue (or purple or green) endpoint, which should persist for at least 20 to 30 seconds. Reweigh the squeeze bottle (final weight).

10. Calculate the amount of juice necessary to supply the recommended daily allowance (RDA) of vitamin C. The report sheet will guide you.

11. Titrate the heated juice sample from step 2 using 10.00 mL of the now cooled juice to prepare the solution as in step 6. Weigh the filled squeeze bottle (initial weight). Titrate the solution with iodine to the blue (or purple or green) endpoint, which should persist for at least 20 to 30 seconds. Reweigh the squeeze bottle (final weight).

12. Pour the vitamin C solution from step 3 off the copper metal. Rinse the copper metal with water and return it to the instructor. Weigh the filled squeeze bottle (initial weight). Titrate the "coppered" vitamin C sample with iodine to the blue endpoint. Reweigh the squeeze bottle (final weight).

13. Pour together the iodide and the vitamin C solutions. Flush all solutions down the drain with plenty of water.

⚠ CAUTION
Acetic acid is corrosive.

EXTRA
Starch and iodine cause a dark blue color, but the dyes in the juice may cause the color to appear purple or green.

EXTRA
1 fl oz = 39.57 mL

TEST FOR VITAMIN C
PRE-LAB QUESTIONS

Name: _____

Lab Partner: _____

Section: _____   Date: _____

1) Since vitamin C is soluble in water, any excess is excreted. Is it better to get all of your RDA at one meal or spread it out over the day? Explain your position.

2) Many soft drinks, snack foods, and energy bars have vitamin C in them. Why do you suppose that is so?

3) If vitamin C is excreted continually, is it possible to get too much?

4) If copper destroys the vitamin C content of foods, would it be prudent to expect that other metals may have the same effect? If so, which metals might you suspect?

TEST FOR VITAMIN C
REPORT SHEET

Name: _____

Lab Partner: _____

Section: _____   Date: _____

I. VITAMIN C CONCENTRATION $\dfrac{250\ \text{mg}}{250\ \text{mL}}$ $=$ 1 mg/mL

II. TITRATION OF VITAMIN C

   a. Initial mass   _____ g

   b. Final mass   − _____ g

   c. Mass of iodine added   _____ g

   d. Grams of iodine per mg vitamin C $= \dfrac{\text{mass of iodine}}{10\ \text{mL vit C}} \times \dfrac{1\ \text{mL vitamin C}}{\text{mg vitamin C}}$

   $$\dfrac{\text{g iodine}}{10\ \text{mL vit C}} \times \dfrac{1\ \text{mL vitamin C}}{\text{mg vitamin C}} = \dfrac{\text{g iodine}}{\text{mg vitamin C}}$$

III. TITRATE JUICE

   a. Type of juice   _____

   b. Initial mass   _____ g

   c. Final mass   − _____ g

   d. Mass of iodine added   _____ g

   e. Mass of vitamin C per mL juice $= \dfrac{\text{mass of iodine}}{10\ \text{mL juice}} \times \dfrac{1\ \text{mg vitamin C}}{_____\ \text{g iodine (II d)}}$

   $$\dfrac{\text{g iodine}}{10\ \text{mL juice}} \times \dfrac{1\ \text{mg vit C}}{_____\ \text{g iodine}} = \dfrac{\text{mg vitamin C}}{\text{mL juice}}$$

   f. Volume of juice to satisfy recommended daily allowance

$$65 \text{ mg vitamin C} \times \frac{\text{mL juice}}{\underline{\hspace{1.5cm}}\text{mg vit C (III e)}} = \underline{\hspace{1.5cm}}\text{mL juice}$$

IV. TITRATION HEATED JUICE

a. Initial mass $\underline{\hspace{3cm}}$ g

b. Final mass   − $\underline{\hspace{3cm}}$ g

c. Mass of iodine added $\underline{\hspace{3cm}}$ g

d. Mass of vitamin C per mL juice = $\dfrac{\text{mass of iodine}}{10 \text{ mL juice}} \times \dfrac{1 \text{ mg vitamin C}}{\underline{\hspace{1.5cm}}\text{g iodine (II d)}}$

$$\frac{\text{g iodine}}{10 \text{ mL juice}} \times \frac{1 \text{ mg vit C}}{\underline{\hspace{1.5cm}}\text{g iodine}} = \frac{\text{mg vitamin C}}{\text{mL juice}}$$

V. QUESTIONS

1. What effect did heating have on the vitamin C in the juice?

2. Would a copper bowl be a wise choice to use when preparing orange juice?

3. How might the results be affected if the vitamin C tablets had been sitting open for a year?

4. Can you think of a way to determine if removing the pulp affects the vitamin C content of juice?

INVESTIGATION

42 *Proteins*

Some Building Blocks

OBJECTIVES

To test various substances for presence of protein structures.

To observe the action of the urease enzyme on a urea solution.

To perform the biuret, xanthoproteic, and calcium oxide tests on proteins.

To discover several processes that denature proteins.

Relates to Chapters 16, 17, and 19 of Chemistry for Changing Times, *twelfth ed.*

BACKGROUND

Proteins are polymers of amino acid units. The bond between amino acid units is an amide bond (nitrogen bonded to the carbon of a carbonyl group C=O) and is sometimes called a *peptide bond*. Proteins are made of many amino acid units. There are, however, only about 20 different amino acids in proteins. The amino acids differ in the R or side-chain group attached to the carbon between the nitrogen and the carbonyl group. When R is a H, for example, the amino acid is glycine, when a CH_3, the amino acid is alanine.

Amino Acid   Protein Chain

The long protein chains also have a particular structure, a distinctive way that the chains are bent and cross-linked to form a three-dimensional shape. When this shape is destroyed, the action of the protein is also destroyed. Called *denaturing*, destroying the structure of a protein can be achieved several ways. Heat, acid, heavy metal ions, and alcohol are some examples of denaturing agents that will be demonstrated in this investigation. Salts and bases such as sodium bicarbonate (baking soda) will also denature proteins.

Proteins are very important parts of the body. In addition to being the main structural part of cells, proteins are also found in skin tissue, fingernails, and hair. Important reactions in the body—such as the action of antibodies and enzymes—are performed by proteins. Saliva contains the enzyme amylase, a protein that begins the digestion of starches while chewing. Another example of the function of an enzyme is the catalysis of the decomposition of urea, $(NH_2)_2CO$, to carbon dioxide (CO_2) and ammonia (NH_3) by the enzyme urease.

Albumin is another protein; one of the main proteins in egg white. When it is denatured, it is no longer soluble in the egg-white fluid. Albumin that is not in solution forms a white opaque semisolid that you would recognize as cooked egg whites or meringue. Casein is the main protein in milk. Denaturing casein forms buttermilk or whipped cream. Gelatin is a protein obtained by boiling animal tissue.

There are several different tests that produce positive results with certain components of proteins. The *biuret test* is positive for multiple amide linkages. Four nitrogen atoms (two on each of two protein strands) complex (share a pair of electrons) with the copper 2^+ ion. This complex of nitrogen atoms and copper causes a pink or purple color in solution.

The *xanthoproteic test* indicates aromatic groups like benzene rings, which are present in tryptophan and tyrosine. Phenylalanine does not give a positive test because of the structure of the amino acid. In this test the aromatic group is nitrated (a NO_2 is attached), generating a yellow color. This phenomenon is also seen when nitric acid gets onto human skin. The skin is temporarily colored yellow (about three days). When a base is added to a protein, the hydrogen of the OH group on the benzene ring of tryptophan or tyrosine is removed, causing an orange or rust color on the surface of the protein.

In another test, ammonia is released from the reaction of protein with calcium oxide. The ammonia comes from the nitrogen end of a protein chain. Each test by itself does not prove that the compound is a protein. However, there are very few non-protein compounds that will produce a positive result for more than one test.

In this investigation, several materials will be tested for protein, and denaturing will be demonstrated.

WASTE AND THE ENVIRONMENT

Only the lead solution is toxic. The lead needs to be converted into an insoluble solid and then buried in a toxic waste site. The metal ion copper is not classified as a toxic metal but in general it is better not to put any metal ions into the water system. Copper ions can be precipitated and buried in a landfill. Calcium compounds are not toxic. The acidic and basic solutions can damage plumbing if they are not neutralized or diluted. The solid proteins are not toxic and are biodegradable.

⚠ *CAUTION*s warn about safety hazards.
*EXTRA*s give helpful hints, additional information, or interesting facts.

Reagents

6 M nitric acid [HNO_3]

6 M sodium hydroxide [NaOH]

calcium oxide [CaO]

1% urease solution

0.1% copper sulfate solution [$CuSO_4$]

phenol red

2% urea solution [H_2NCONH_2]

concentrated nitric acid [HNO_3]

ethyl alcohol [CH_3CH_2OH] ethanol

3 M sodium hydroxide [NaOH]

1 M lead nitrate [$Pb(NO_3)_2$]

1 M acetic acid [CH_3COOH]

sodium sulfide [Na_2S]

sodium carbonate [Na_2CO_3]

1 M iron(III) chloride [$FeCl_3$]

distilled water

Common Materials

egg

unflavored gelatin

nail clippers

milk

hair

Laboratory Equipment

red litmus

pH paper

filter paper

several large test tubes

3 250 mL beakers

ring stand, ring and wire gauze

laboratory burner

watch glass

dropper

10 mL graduated cylinder

338

PROCEDURE

One pair of students prepare albumin; another pair prepare gelatin and share solutions.

1. **Albumin Solution**

   Separate 1 egg white from the yolk by pouring the yolk back and forth from 1 shell half to the other above a 250 mL beaker. Add 200 mL of distilled water. Stir gently for 30 seconds. Set the solution aside for 5 minutes. Take samples from the middle with a pipet, avoiding the floating material on top and the residue on the bottom.

   ⚠ *CAUTION*
   Use a safety bulb with a pipet.

2. **Gelatin Solution**

   Heat 250 mL of distilled water to boiling. Remove the flame. With stirring, add a package of unflavored gelatin. Stir the solution until all of the gelatin is dissolved. Set aside to cool.

   ⚠ *CAUTION*
   Don't burn yourself.

3. **Xanthoproteic Test on a Fingernail**

   One student of each pair should clip off a small piece of fingernail. Place the fingernail on a watch glass. Add 1 or 2 drops of 6 M nitric acid (HNO_3). If protein is present, a yellow color should appear. Add several drops of 6 M sodium hydroxide (NaOH). The orange color is a positive test for protein.

   ⚠ *CAUTION*
   Nitric acid and sodium hydroxide are caustic.

   EXTRA
   Bring nail clippers.

4. **Test on Hair**

   Place a small wad of hair in a test tube. Add 1 mL of distilled water. Add a small scoop of calcium oxide (CaO). Heat gently. The smell of ammonia is one positive test. Place a piece of moist red litmus paper at the mouth of the test tube. The ammonia will cause a color change in the litmus paper.

   EXTRA
   Collect the hair from a hairbrush.

5. **Biuret Test on Hair**

   Place a small wad of hair in a test tube. Add 1 mL of water. Add 1 mL of 6-M sodium hydroxide. Add several drops of 0.1% copper sulfate ($CuSO_4$) solution. A pale pink or violet color is a positive test. The test tube may have to sit for a time for the color to develop. View the color against a white sheet.

   ⚠ *CAUTION*
   Sodium hydroxide is caustic.

6. **Biuret Test on Saliva**

   Place several drops of saliva or a commercial amylase sample in a test tube. Add several drops of 6-M sodium hydroxide. Shake gently to mix. Add several drops of 0.1% copper sulfate solution. A pale pink or violet color is a positive test. The color is easier to see against a white sheet.

   ⚠ *CAUTION*
   Sodium hydroxide is caustic.

7. **Biuret Test on an Enzyme**

   Place 2 mL of 1% urease solution in a test tube. Add 1 mL of 6 M sodium hydroxide. Shake gently to mix. Add several drops of 0.1% copper sulfate solution. A pale pink or violet color is a positive test. The color is easier to see against a white sheet.

⚠ *CAUTION*
Sodium hydroxide is caustic.

8. **Enzyme Action**

   Place 2 mL of 1% urease solution in a test tube. Add 3 drops of phenol red indicator to the test tube. Add 5 mL of 2% urea solution. Shake gently to mix. Watch for a color change.

EXTRA
Phenol red is yellow in acidic solutions and red in basic solutions. Ammonia is a base.

9. **Biuret Test**

   Test albumin, gelatin, and casein with the biuret test by adding 1 mL of 6 M sodium hydroxide to 2 mL of each solution in a test tube. Shake gently to mix. Add a drop at a time of 0.1% copper sulfate until a light pink or violet color appears. Record your observations.

EXTRA
Casein is the main protein in milk. Use milk for the tests.

10. **Xanthoproteic Test**

    Test gelatin, albumin, and casein by the xanthoproteic test by adding 1 mL of concentrated nitric acid to 2 mL of each solution in a test tube. Mix. Warm the mixtures carefully. The solutions will be yellow if a protein is present. Cool the test tubes with water. Slowly add several milliliters of 3 M sodium hydroxide to each tube. The appearance of an orange color is a positive test. Record your observations.

⚠ *CAUTION*
Nitric acid is corrosive.

⚠ *CAUTION*
Sodium hydroxide is caustic.

11. **Denaturing by Heat**

    Place 5 mL of albumin, casein, and gelatin in three separate test tubes. Label them. Heat the upper portion of each solution gently until it boils. Set the casein solution aside until it cools. The scum on top is denatured casein. Albumin in urine is detected by this method. Record your observations.

⚠ *CAUTION*
Don't burn yourself. Don't point the test tube at anyone.

12. **Denaturing by Alcohol**

    Place 1 mL of each protein solution (albumin, gelatin, and casein) in a separate test tube. Add 1 mL of ethyl alcohol to each test tube. Record your observations. It may help to tilt the test tubes to see the small curds formed.

13. **Denaturing by Pb(NO₃)₂**

   Place 1 mL of each protein solution in a separate test tube. Add several drops of 1 M Pb(NO₃)₂ to each. Record your observations.

14. **Denaturing by Acid**

   Place 1 mL of each protein solution in a separate test tube. Add 1 mL of 1 M acetic acid. Record your observations.

15. Collect all used solutions containing lead. Add a threefold excess of sodium sulfide (Na₂S). Let the solution sit for an hour. Adjust the pH of the solution to neutral (pH 7 or more). Filter the precipitate. Discard the precipitate and filter in a container to be buried in a secure toxic-waste landfill. The filtrate contains sulfide. Remove the sulfide by adding 1-M iron (III) chloride (ferric chloride) in a three-fold excess. Neutralize the solution with sodium carbonate (Na₂CO₃). Pour the solution through filter paper prepared as shown below.

EXTRA
A tablespoon of vinegar in a cup of milk makes "buttermilk."

EXTRA
Use pH paper to be sure that the solution is not acidic.

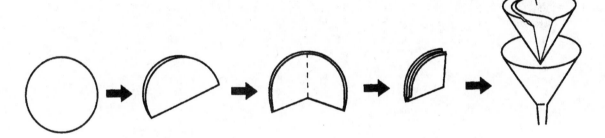

   Put the precipitate and filter in a container to be buried in a toxic-waste landfill. Flush the filtrate down the drain with plenty of water.

16. Evaporate the copper solutions and place the residue in a container to be buried in a landfill. Throw the protein solid in the trash, and flush the solutions down the drain with plenty of water.

PROTEINS
PRE-LAB QUESTIONS

Name: _____

Lab Partner: _____

Section: _____   Date: _____

1) Why is it important to keep the pH levels in the body constantly at the "optimum" pH for protein function?

2) Why are we very careful to keep lead and other metals out of the water systems following lab investigations?

3) What is the possible result of alcohol consumption on enzymes and proteins in the digestive system? In the circulatory system?

4) List as many body parts as you can that are made of protein—for example, hair. Which of these are most in danger of being denatured? Which would cause the most serious effect on the physiology of a human if it were denatured?

PROTEINS
REPORT SHEET

Name: _____

Lab Partner: _____

Section: _____   Date: _____

I. XANTHOPROTEIC TEST ON NAIL   Positive _____   Negative _____

II. HAIR
   1. Test for ammonia   Positive _____   Negative _____

   2. Biuret Test   Positive _____   Negative _____

III. SALIVA
   1. Biuret Test   Positive _____   Negative _____

IV. ENZYME
   1. Biuret Test   Positive _____   Negative _____

   2. Test for enzyme action (ammonia)   Positive _____   Negative _____

V. PROTEIN TESTS

	Albumin	Gelatin	Casein
Biuret	_____	_____	_____
Xanthoproteic	_____	_____	_____

VI. DENATURING

	Albumin	Gelatin	Casein
Heat	_____	_____	_____
Alcohol	_____	_____	_____
$Pb(NO_3)_2$	_____	_____	_____
Acid	_____	_____	_____

VII. QUESTIONS

1. If a person swallowed a heavy metal, would it be useful to drink milk and/or egg white? Explain.

2. Why is alcohol used as a disinfectant?

3. What causes cooked egg to be white?

4. What caused the phenol red to change color in step 8?

5. Urease is a specific enzyme. What is meant by that?

6. What else might be added to the list of denaturing agents in Section V?

7. Describe the differences in texture and appearance of the denatured solutions in steps 11, 12, 13, and 14.

8. Enzymes and other vital biochemical molecules are proteins. What would be the result of denaturing one of these proteins?

43 *Enzymes*

Biological Catalysts

OBJECTIVES

To prepare a catalase solution in the form of potato extract.

To observe the action of the enzyme catalase on hydrogen peroxide.

To measure the amount of oxygen produced from the decomposition hydrogen peroxide by catalase.

To introduce various denaturing factors and measure the effects on the enzyme action of catalase.

Relates to Chapters 16 and 17 of Chemistry for Changing Times, *twelfth ed.*

BACKGROUND

Enzymes are biological *catalysts* that allow reactions to occur at a lower temperature than without the enzyme. Catalysts (enzymes) are not consumed in a reaction, but remain unchanged at the end of the reaction, so that they are available to be used over and over again. The human body makes use of many enzymes. Human cells each contain about 2000 enzymes that catalyze 100 or more reactions per minute. Without enzymes the reactions would occur slowly at a temperature as low as 37°C (body temperature).

Examples of enzymes that are found in our bodies include amylase in saliva that begins the breakdown of starch into glucose units. Ptyalin is a salivary amylase. It begins the breakdown of starches in the mouth. Sucrose, a disaccharide, is broken into two monosaccharides, glucose and fructose, by the action of sucrase. The enzyme pepsin aids in breaking down certain proteins in the stomach. Thrombin is the blood-clotting enzyme. After the tissue is healed, the clot is dissolved by the catalyst plasmin. The list of enzymes goes on and on.

A common enzyme is catalase. Catalase is present in many plant and animal tissues including potatoes. One reaction catalyzed by catalase is the decomposition of hydrogen peroxide (H_2O_2) to oxygen and water as shown in the following equation.

$$2H_2O_2 \xrightarrow{\text{Catalase}} 2H_2O + O_2$$

Several factors affect enzyme activity. Catalase contains iron, and certain ions such as cyanide and sulfide inhibit the enzyme activity by *complexing* with the iron. (Complexing is a process in which ions in solution form bonds with other ions or molecules by sharing lone pairs of electrons.) Also, because enzymes are most active in specific temperature ranges, many enzymes are destroyed by elevated temperatures. Furthermore, enzymes generally operate in a narrow pH range so that the presence of acids and bases can reduce their effectiveness. Another way to destroy an enzyme is with the presence of a heavy metal such as lead.

Most enzymes are proteins. The same things that *denature* (destroy the structure of) proteins also denature enzymes, or at least inhibit their activity. Heavy metals, high temperature, acids, bases, and agitation all denature proteins. In this investigation, you will observe the action of catalase upon the decomposition of hydrogen peroxide. You will also alter the activity of catalase using several destructive agents.

WASTE AND THE ENVIRONMENT

Lead ions, which are classified as toxic, and sulfide ions should not be introduced into the water system. Acidic and basic solutions may damage plumbing if not neutralized or diluted. The potato pieces, extract, and hydrogen peroxide are not toxic and are biodegradable.

⚠ *CAUTION*s warn about safety hazards.

*EXTRA*s give helpful hints, additional information, or interesting facts.

<u>Reagents</u>

3% hydrogen peroxide [H_2O_2]	6 M hydrochloric acid [HCl]
3 M sodium sulfide [Na_2S]	6 M sodium hydroxide [NaOH]
1 M lead(II) nitrate [$Pb(NO_3)_2$]	1 M iron(III) chloride [$FeCl_3$]
sodium carbonate	

<u>Common Materials</u>

potato, freshly peeled   paring knife

<u>Laboratory Equipment</u>

ring stand, ring, and wire gauze	15 cm test tube
2 250 mL beakers	1-hole stopper with glass tube to fit test tube
600 mL beaker	pH paper
2 150 mL beaker	stopwatches
1000 mL beaker	utility clamp
100 mL graduated cylinder	rubber tubing, 2 ft.
10 mL graduated cylinder	thermometer

PROCEDURE

1. Cut a 30 g portion of a clean peeled potato into cubes that are 5 mm on a side. Place the cubes in a 150 mL beaker with 50 mL of water and let it stand for 10 minutes. Swirl the water occasionally.

EXTRA
It is important that the potato is in small pieces. Mashing with a mortar and pestle will also help.

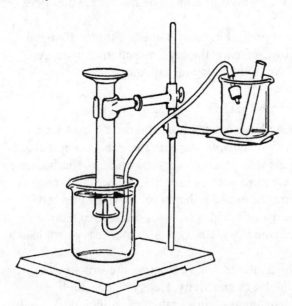

2. Set up the gas-measuring apparatus shown in the figure above by filling a 1000 mL beaker with about 800 mL of water. Fill a 100 mL graduated cylinder with water. Place your fingers or palm tightly over the top and invert the cylinder into the beaker, letting as little air into the cylinder as possible. Then clamp the cylinder to a ring stand. Insert the rubber tubing into the cylinder so that any gas passing through the tubing will go into the cylinder. (Keep the stopper end elevated or the hose will act like a siphon and will allow the water to drain out of the 1000 mL beaker through the hose.)

3. Set a 600 mL beaker on an elevated ring attached to the ring stand. Place a dry 15-cm test tube in the beaker.

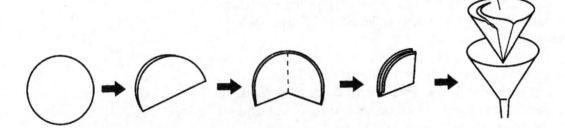

4. Filter the potato solution and allow the filtrate to collect in a 150 mL beaker. This is the potato extract you will use.

5. Read and record the water level in the 100 mL cylinder as the initial reading.

EXTRA
Zero is at the top.

6. Set up two hot-water baths in 250 mL beakers: one at 37°C, one at 70°C. Place 5 mL of potato extract in a test tube in each bath. Allow it to sit for 10 minutes.

⚠ *CAUTION*
Don't burn yourself.

7. Place 5 mL of the room-temperature potato extract in the 15-cm test tube. Add 6 mL of 3% hydrogen peroxide and gently shake the tube to mix completely. Proceed quickly to step 8.

EXTRA
The H_2O_2 needs to be fresh.

8. When one student says "Start," the other student should stopper the test tube quickly and tightly with the stopper attached to the tube in the graduated cylinder. Allow the oxygen to collect until 8 mL has been collected. Record the time, unplug the test tube, and record the amount of oxygen collected.

9. Refill the graduated cylinder and repeat steps 7 and 8, collecting oxygen for the same length of time using the same

amount of potato extract, but add two drops of 3-M sodium sulfide. Add 6 mL of 3% hydrogen peroxide. Mix well.

10. Refill the graduated cylinder and repeat steps 7 and 8, collecting oxygen for the same length of time using the same amount of potato extract, but add 20 drops of 1 M lead nitrate and 6 mL of 3% hydrogen peroxide. Mix well.

11. Refill the graduated cylinder and repeat steps 7 and 8, collecting oxygen for the same length of time using the same amount of 37°C potato extract and 6 mL of 3% hydrogen peroxide. Mix well.

12. Refill the graduated cylinder and repeat steps 7 and 8, collecting oxygen for the same length of time using the same amount of 70°C potato extract and 6 mL of 3% hydrogen peroxide. Mix well.

13. Refill the graduated cylinder and repeat steps 7 and 8, collecting oxygen for the same length of time using 5 mL of original potato extract and 2 drops of 6 M HCl. Add 6 mL of 3% hydrogen peroxide. Mix well.

14. Refill the graduated cylinder and repeat steps 7 and 8, collecting oxygen for the same length of time using 5 mL of original potato extract and 2 drops of 6 M NaOH. Add 6 mL of 3% hydrogen peroxide. Mix well.

EXTRA
Use pH paper to ensure that the solution is __not__ acidic. Sodium sulfide in acid solution produces poisonous hydrogen sulfide gas.

15. Pour the lead-potato extract and the sulfide-potato extract together. Add 10 drops of 3 M sodium sulfide. Let the solution sit for an hour. Adjust the pH of the solution to neutral (pH 7 or more). Filter the precipitate. Discard the precipitate and filter in a container to be buried in a secure toxic waste landfill. The filtrate contains sulfide. Remove the sulfide by adding 1 M iron(III) chloride (ferric chloride, $FeCl_3$) in a threefold excess. Neutralize the solution with sodium carbonate (Na_2CO_3). Pour the solution through filter paper prepared as shown below. Put the precipitate and filter in a container to be buried in a toxic waste landfill. You may flush the filtrate down the drain with plenty of water. Throw the solid pieces of potato in the trash. Flush the solutions down the drain with plenty of water.

ENZYMES
PRE-LAB QUESTIONS

Name: _____

Lab Partner: _____

Section: _____   Date: _____

1) Why should it be easier to get the extract from the potato if it is in smaller pieces?

2) If cyanide and sulfide ions will inhibit the function of catalase because it contains iron, what other important protein function might be disrupted by the presence of these ions?

3) Our bodies have elaborate systems for maintaining optimum temperature and pH levels. Why?

4) Why is it important to use the same amount of catalase and hydrogen peroxide in each trial?

ENZYMES
REPORT SHEET

Name: _____

Lab Partner: _____

Section: _____ Date: _____

I. CATALASE

Final water level _____ mL
Minus initial water level − _____ mL
Oxygen produced _____ mL
Time elapsed _____ sec

II. CATALASE + SULFIDE

Final water level _____ mL
Minus initial water level − _____ mL
Oxygen produced _____ mL

III. CATALASE + LEAD

Final water level _____ mL
Minus initial water level − _____ mL
Oxygen produced _____ mL

IV. CATALASE AT 37ºC

Final water level _____ mL
Minus initial water level − _____ mL
Oxygen produced _____ mL

V. CATALASE AT 70ºC

Final water level _____ mL
Minus initial water level − _____ mL
Oxygen produced _____ mL

VI.   CATALASE IN ACID

Final water level   _____ mL
Minus initial water level   − _____ mL
Oxygen produced   _____ mL

VII. CATALASE IN BASE

Final water level   _____ mL
Minus initial water level   − _____ mL
Oxygen produced   _____ mL

VIII.   QUESTIONS

1. What effect did the sulfide have?

2. What effect did the lead have?

3. At which temperature does catalase work best: 25°C, 37°C, or 70°C?

4. At which temperature does catalase work the least well?

5. How does the acidity affect the enzyme action?

44 *Forensic Chemistry*

You Always Leave a Trace!

OBJECTIVES

To gain an understanding of practical uses for chemical processes.
To produce a chemiluminescent reaction showing the presence of blood.
To show the detection of blood using a chemical reaction.
To illustrate the collection of a plastic and a latent fingerprint.

Relates to Chapter 16 of Chemistry for Changing Times, *twelfth ed.*

BACKGROUND

Forensic chemistry is the use of analytical chemistry to evaluate evidence and determine exactly what happened at a crime scene. A forensic chemist uses chemical processes in everyday investigations that help identify suspects, determine actions, sequences of events, timeframes, mechanisms of the crime, etc. This assessment is accomplished through fingerprinting, DNA analysis, examination of footprints and blood spatters, tire tracks, pieces of glass or other materials from broken objects, even analysis of cigarette ashes. Processes that use clothing fibers, pollen, and spores to place a person at a crime scene are part of a recently developed science called *palynology*.

Two of the most conclusive types of evidence at a crime scene are fingerprints and blood. Both of these can lead to specific identities of the people involved at the scene. Fingerprints are classified as either latent or plastic. *Latent prints* are generated because some oils of the hand are left on a smooth surface. *Plastic prints* are ones left in soft materials, such as mud or paint, and are three-dimensional in nature. The soft material has formed to the skin filling the hills and valleys on the finger. Fingerprints are easily destroyed and latent prints may have a limit to the time in which they remain useful. Plastic prints can dry and remain viable for a much longer time. Plastic prints of shoe soles and tires can last for months or years if they remain undisturbed.

There are two steps to gathering fingerprint evidence from a crime scene. The print must first be detected. This is accomplished by dusting it with a very fine powder such as graphite or cupric oxide or by a vapor method. Once the print is located, it must be "lifted," or retrieved and preserved for later comparison.

Blood evidence is not always visible to the naked eye. Even after a surface has been washed free of visible blood, a faint residue can be detected with chemical reactions that make use of the catalytic nature of the iron ions in hemoglobin. The Kastle-Meyer phenolphthalein test for blood works because the iron (II) ions in hemoglobin *catalyze* the hydrogen peroxide oxidation of phenolphthalein to its pink state. Another test for blood uses *luminol,* which is oxidized by hydrogen peroxide to a compound that will emit a greenish light. This process is called *chemoluminescence*. Ultraviolet light or "black light" will also cause the luminol-blood combination to glow more brightly. Only a miniscule amount of blood is required for the reaction because the iron is only a catalyst; therefore, these tests are extremely sensitive. The persistence of blood stains caused Lady MacBeth to say "Out, *@#%ed spot! Out I say!"

WASTE AND THE ENVIRONMENT

The zinc solution needs to be evaporated in a metal pan because metal ions should not be put into the water system. The other solutions are not toxic if diluted with water.

⚠ *CAUTION*s warn about safety hazards.
*EXTRA*s give helpful hints, additional information, or interesting facts.

Reagents
    luminol
    3% hydrogen peroxide [H_2O_2]
    ethyl alcohol [CH_3CH_2OH]
    phenolphthalein
    sodium hydroxide [NaOH]
    zinc powder
    copper oxide [CuO] fine powder

Common Materials
    clear tape
    ultraviolet light
    cotton swabs or Q-Tips
    modeling clay or Handi-Tak® or soft putty substance
    blood stains

Laboratory Equipment
    watch glass
    eye dropper
    150 mL beaker

Special Equipment
    A "dark box" can be made by cutting holes in a cardboard box. One hole will allow the UV light in and the other will let the student see the sample.

PROCEDURE

The solution for the Kastle-Meyer phenolphthalein test is prepared by mixing 0.25 g of phenolphthalein, 2.5 g of sodium hydroxide (NaOH), 2.5 g of zinc powder, and 32.5 mL of distilled water. Gently heat the mixture for 1 hour in a beaker with a watch glass over it.

Part A: FINGERPRINT DETECTION
Latent Print
1. Clean a watch glass. Carefully place a thumbprint in the middle of the watch glass. With an eyedropper, draw up a small amount of powdered cupric oxide (CuO). Squirt the CuO on the fingerprint. Shake the watch glass so the CuO covers the fingerprint. Gently pour the excess CuO onto a piece of paper. Using the paper as a funnel, return the excess CuO to the bottle of CuO.

2. Carefully place a piece of clear tape on top of the fingerprint. Rub the tape so that it adheres to the fingerprint. Gently lift the tape and place it on your Report Sheet.

Plastic Print
3. Take a piece of modeling clay or Handi-Tak® the size of a pencil eraser. Roll it in your hands until it is soft. Mash it flat on the bottom of a beaker with your thumb. If you did not wiggle your thumb, the clay now contains a plastic fingerprint. Gently pull up the clay. Tape the fingerprint to the report form.

Part B: BLOOD DETECTION
Luminol Test
4. Mix 0.1 g of luminol with 30 mL of 3% hydrogen peroxide in a spray bottle. Spray the suspected area with the solution.

5. When the solution has dried, place the sample in a dark box. Shine a ultraviolet light on the area. The area that glows with a yellow-green color has been in contact with blood.

Kastle-Meyer Phenolphthalein Test
6. Moisten a cotton swab (or Q-Tip) with distilled water. Wipe suspected stain with the swab. Put a drop of ethyl alcohol (CH_3CH_2OH) on the swab.

7. Shake the bottle of phenolphthalein-zinc solution. Add one drop to the swab.

⚠ *CAUTION*
Don't burn yourself.

EXTRA
Don't smudge the print.

EXTRA
The hydrogen peroxide needs to be fresh.

8. Add one drop of hydrogen peroxide (H_2O_2) to the swab. If the stain is blood, the swab will turn bright pink if the stain is blood.

   EXTRA
   The hydrogen peroxide
   must be fresh.

9. The phenolphthalein-zinc solution should be allowed to evaporate on a metal pan. The zinc oxide dust produced can then be put in a waste container to be buried. The other solutions can be flushed down the drain with lots of water.

FORENSIC CHEMISTRY
PRE-LAB QUESTIONS

Name: _____

Lab Partner: _____

Section: _____   Date: _____

1) When might a latent fingerprint be permanent?

2) If iron(II) ion is the catalyst for tests that detect the presence of blood, what compounds might give a false positive test?

3) Will the tests for blood work only for human blood or will they detect any animal blood?

4) What characteristics must be possessed by the powder used to dust for fingerprints?

FORENSIC CHEMISTRY
REPORT SHEET

Name: _____

Lab Partner: _____

Section: _____   Date: _____

I. PART A: FINGERPRINT DETECTION
    Latent Print

    Plastic Print   Instructor's Initials _____

II. PART B: BLOOD DETECTION
    Luminol Test

SAMPLE	POSITIVE	NEGATIVE
_____	_____	_____
_____	_____	_____
_____	_____	_____

    Kastle-Meyer Phenolphthalein Test

SAMPLE	POSITIVE	NEGATIVE
_____	_____	_____
_____	_____	_____
_____	_____	_____

III. QUESTIONS

1. Why does a latent print show up better on a smooth surface?

2. In the luminol test, the iron(II) ion in hemoglobin is acting as a catalyst. Other catalysts also work such as iodine, permanganate, or hypochlorite (bleach). What would happen if the suspected area had one of the other catalysts spilled on it?

3. Why is a dark room better for observing the luminol test?

45 *Protein in Milk and Other Foods*

We Drink That Solid?

OBJECTIVES

To isolate the protein casein from milk by coagulation with an acid.
To perform biuret and xanthroproteic protein tests on the casein.
To test several foods for evidence of protein.

Relates to Chapters 17 and 19 of Chemistry for Changing Times, *twelfth ed.*

BACKGROUND

Milk is a complex mixture of several components, including proteins and fats. The main protein is casein. Like other proteins, casein can be coagulated by several methods. One way is to heat it. The skin that forms on heated milk is coagulated protein. Agitation also will cause coagulation, as in the process of making whipped cream. Proteins will also coagulate in an acid solution. A substitute for buttermilk as an ingredient in a recipe can be made by adding a tablespoon of vinegar or lemon juice to a cup of milk.

Cheese is made from milk proteins that have been concentrated. The casein separated in this investigation is ricotta cheese and could be made into aged cheese such as cheddar or Swiss. Ricotta cheese is the cheese you put in lasagna and stuffed pasta. Casein can also be made into paints, glues, and some plastics.

Proteins are a major component of our bodies. Proteins are used as structural cell material, and as enzymes and antibodies. Proteins also show up in paper coatings, pharmaceuticals, and textile fibers. Our body makes its proteins from *amino acids* that are taken from the protein we eat. (We are what we eat.)

Many foods contain protein. Some foods such as meat, dairy foods, and other animal products contain what is called *complete protein*. That is, all the amino acids necessary for the body are provided in that protein source. Other proteins such as those from plants lack one or more of the necessary amino acids. Luckily, a combination of a legume (beans, peas, peanuts) and a grain (corn, oats wheat, rye) make a complete protein because one class of foods has the amino acid that the other is missing. It is no accident that most traditional ethnic foods (Cajun red beans and rice, Italian pasta and tomato sauce, Southern black-eyed peas and cornbread) are combinations that form inexpensive but complete protein sources.

We call the bond between the C and N joining two amino acids a *peptide linkage*. The peptide linkage forms an *amide group*:

$$\begin{array}{ccc} O & & H \\ \parallel & & \mid \\ -\!\!-C & -\!\!-N & -\!\!- \end{array}$$

Amide

Many amino acids linked together in this manner form a protein.

Two reactions that can be used to test for protein in foods are the xanthoproteic test and the biuret test. The xanthoproteic test produces an intense yellow color by *nitrating* (adding a NO_2 group) the aromatic ring in certain amino acid side chains (tyrosine and tryptophan). Phenylalanine does not produce a positive xanthoproteic test because of its unique structure. The xanthoproteic phenomenon is also seen when nitric acid gets onto human skin. The skin is temporarily (three days) colored yellow. The addition of base removes the hydrogen from the OH group on the aromatic ring in tryptophan or tyrosine. This causes a shift in the energy of the electrons in the compound. The new color of orange or rust is a confirmation of the presence of at least one of the two amino acid groups that are a part of most proteins.

A positive biuret test indicates the presence of two amide groups joined by a single carbon or nitrogen. Four nitrogens (two on each of two proteins strands) complex (share a pair of electrons) with the copper 2+ ion. The complex causes a purple or pink color in solution.

WASTE AND THE ENVIRONMENT

The solutions, foods, and milk casein are not toxic. Acids and bases can cause damage to plumbing if not diluted or neutralized. Although normally it is not good procedure to put metal ions such as copper into the water system, in this investigation the amount is so small that it presents little risk to the environment. Two or three drops of a 1% solution diluted with plenty of water quickly become a negligible concentration.

⚠ *CAUTION*s warn about safety hazards.
*EXTRA*s give helpful hints, additional information, or interesting facts.

Reagents

glacial acetic acid [CH_3COOH]	concentrated ammonia [NH_3]
1:1 ethyl alcohol-ethyl ether	0.1% M copper sulfate [$CuSO_4$]
concentrated nitric acid [HNO_3]	6 M sodium hydroxide [$NaOH$]

Common Materials

milk	white meat (turkey or chicken)
various foods to test for protein	marshmallow
rice	radish
white bread	sugar
butter beans	sliced almonds
Swiss cheese	cheesecloth, 12 in. square
rubber bands (to fit a 500 mL beaker)	

Laboratory Equipment

600 mL beaker	filter paper
100 mL beaker	2 filter flasks
125 mL Erlenmeyer flask	ring stand, ring and wire gauze
watch glass	laboratory burner
thermometer	eye droppers

PROCEDURE

1. Prepare a hot water bath in a 600 mL beaker containing about 350 mL of water. Place the beaker on a ring stand and heat the water with a Bunsen burner or hotplate, if available. Place 30 g of milk in a 125 mL Erlenmeyer flask. Heat the sample in the water bath until it reaches about 40°C.

   ⚠ *CAUTION*
   Don't burn yourself.

2. While swirling the milk sample, add **8** drops of glacial acetic acid (CH_3COOH). Continue to add drops of acetic acid as long as casein continues to precipitate.

   ⚠ *CAUTION*
   Acetic acid is corrosive.

3. Placing several layers of cheesecloth (or gauze bandage material) across the top of a 600 mL beaker and hold the cloth in place with a rubber band. Filter the sample through cheesecloth by slowly pouring the coagulated milk sample through the cloth. While wearing gloves, wrap the cheesecloth around the solid (curd) that is separated from the liquid. Squeeze out most of the liquid. Save the liquid (whey) to test for protein later.

   EXTRA
   Vinegar is dilute acetic acid.

4. Place the solid protein and fat (the curd) in a 100 mL beaker. Add 30 mL of a 1:1 mixture of ethyl alcohol-ethyl ether to dissolve the fat. Stir the mixture well and filter by suction through filter paper in a Buchner funnel. Rinse the beaker into the filter with small portions of the 1:1 solvent mixture to remove all the casein from the beaker.

   ⚠ *CAUTION*
   The alcohol-ether mixture is very flammable.

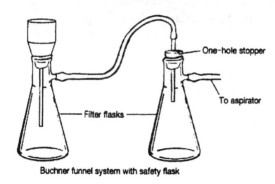

Buchner funnel system with safety flask

5. Air dry the protein by continuing the suction for several minutes.

6. Weigh a watch glass and record the mass. Scrape the casein off the filter paper onto the watch glass and reweigh it.

7. Test the casein by both protein tests as outlined in steps 8 and 9.

8. **Xanthroproteic Test**
    Place a small amount (about a teaspoon) of food material on a watch glass. Place a drop of concentrated nitric acid (HNO_3) on the material. If protein is present, a yellow color will appear. Rinse off the nitric acid with water. Add a drop of concentrated ammonia (NH_3) solution to the spot. An orange color indicates the presence of protein.

    ⚠ *CAUTION*
    Nitric acid and ammonia are both corrosive.

9. **Biuret Test**
    Add a few drops of 6 M sodium hydroxide (NaOH) to a sample of food on a watch glass, then add a few drops of dilute (0.1%) copper sulfate ($CuSO_4$) solution. The reddish violet to violet-blue color indicates a peptide linkage, which is present in all proteins.

    ⚠ *CAUTION*
    Sodium hydroxide is caustic.

10. Evaporate two small samples of whey on watch glasses by placing them one at a time on top of a beaker that contains boiling water. Test the samples of whey by both protein tests.

11. Test a small amount (less than a gram) of the following foods by the protein tests: rice, butter beans, bread, sliced almond, Swiss cheese, and white meat (chicken or turkey). To show that everything is not protein, test a marshmallow, radish, and sugar. Several samples can be placed on the same watch glass and tested at the same time. Some of the tests require several minutes to develop.

    EXTRA
    Split the bean in half. Test the inside. Test the white part of the radish and bread.

12. Throw the milk protein and foods into the wastebasket. Wash the whey and other solutions down the sink with plenty of water. Flush the acid down the drain with an excess of water.

PROTEIN IN MILK AND OTHER FOODS
PRE-LAB QUESTIONS

Name: _____

Lab Partner: _____

Section: _____   Date: _____

1) Enzymes, and other vital biochemical molecules are proteins. What would be the result of eating too little protein?

2) What might you expect to be the visible result if you accidentally splattered sodium hydroxide and copper sulfate on your skin at the same time?

3) Why will we test the inside of the butter bean and the white portion of the bread and radish?

PROTEIN IN MILK AND OTHER FOODS
REPORT SHEET

Name: _____

Lab Partner: _____

Section: _____   Date: _____

I. CASEIN IN MILK

Mass of watch glass and casein _____ g

Minus mass of watch glass −_____ g

Mass of casein _____ g

The percentage of casein in the milk sample $= \dfrac{\text{mass of casein}}{\text{mass of milk}} \times 100\%$

$$= \dfrac{\text{g casein}}{30\,\text{g milk}} \times 100\% = \underline{\hspace{2cm}} \%$$

Brand name of milk _____

Percentage of protein listed on the product label _____

II. PROTEIN TESTS
Indicate the color of the sample after the addition of each reagent

Sample	HNO_3 (conc.)	Water + NH_3	$CuSO_4$ + NaOH	Conclusion
Casein				
Whey				
White meat				
Radish				
Marshmallow				
Rice				
Butter beans				
Bread				
Sliced almond				
Swiss cheese				
Sugar				

III. QUESTIONS

1. Did you separate all the protein from the whey? Use your results to support your answer.

2. List as many foods as possible that contain protein.

3. Were the two protein tests consistent in their results? Explain.

46 *Production of Carbon Dioxide*

CO₂ Gas from Everywhere!

OBJECTIVES

To measure the amount of carbon dioxide released from several common products using the technique of water displacement.

To calculate the amount of carbon dioxide gas released per gram of solid.

To understand the reactions that lead to the release of carbon dioxide from sodium bicarbonate and calcium carbonate.

Relates to Chapters 17 and 5 of Chemistry for Changing Times, *twelfth ed.*

BACKGROUND

Carbon dioxide gas is produced in many ordinary reactions. Some of the reactions we will observe in this investigation involve sodium bicarbonate, a compound found in many dietary applications. The bicarbonate ion reacts in water to produce carbon dioxide gas in a two-step reaction. The bicarbonate ion (HCO_3^-) reacts with a hydronium ion in solution to produce the unstable compound carbonic acid (H_2CO_3) and a water molecule. *Hydronium ions*, H_3O^+, are hydrogen ions (H^+) bound to water molecules and are the result of the interaction of a H^+ often from an acid, reacting with a molecule of water. A bicarbonate ion takes the hydrogen ion from the hydronium ion, leaving a water molecule and producing carbonic acid. Carbonic acid immediately dissociates into water and carbon dioxide gas.

$$HCO_3^- + H_3O^+ \rightarrow H_2O + H_2CO_3$$
$$H_2CO_3 \rightarrow H_2O + CO_2.$$

This reaction is the basis for common phenomena such as the rising of bread dough and cake batter, the fizzing of an Alka-Seltzer tablet in water, and the neutralization of excess acid in the stomach by some antacids.

The bicarbonate ion is often found combined with a sodium ion as sodium bicarbonate ($NaHCO_3$). Grocery stores sell this compound as baking soda. It is also known as bicarbonate of soda. In recipes using baking soda, a compound containing an acid, such as lemon juice (citric acid), vinegar (acetic acid), or buttermilk (lactic acid) must be added as an ingredient so the chemical reaction releasing carbon dioxide (CO_2) will occur.

Baking powder is a mixture of baking soda, filler, and acids in dry powder form. As the acids dissolve in water, they react with the sodium bicarbonate to produce carbon dioxide. One of the acids is soluble in water at room temperature. The other acid dissolves only slightly at room temperature but is soluble in hot water. Baking powder used in a recipe does not require an extra source of acid, but part of the carbon dioxide is not produced until the ingredients of the recipe is heated. Shortly after the ingredients of a recipe are mixed, part of the carbon dioxide is produced by the reaction of the sodium bicarbonate and the acid that is soluble in water at room temperature. After baking begins, more carbon dioxide is produced when the temperature rises to the point at which the other acid becomes soluble and reacts with sodium bicarbonate.

Baking soda and baking powder are both used extensively in cooking, especially in baking. The carbon dioxide generated by baking powder and baking soda is responsible for most of the open spaces in bread, cakes, and other baked goods. The spaces are generated as the dough rises. When the dough is baked, the structure formed by the CO_2 gas in the dough becomes solid. Most of the carbon dioxide escapes, leaving the open spaces. These spaces prevent the baked goods from being dense and heavy as in unleavened bread.

Carbon dioxide is also generated by antacids. Alka-Seltzer® contains aspirin (a pain reliever), sodium bicarbonate, and citric acid. When the Alka-Seltzer® tablet is dissolved in water, the citric acid in water produces hydronium ions. The hydronium ions react with the bicarbonate part of sodium bicarbonate to produce carbonic acid, which immediately produces water and carbon dioxide.

As the solution is consumed, some of the carbon dioxide helps to relieve stomach distress by inducing belching. (Luckily for the consumers and those around them, most of the carbon dioxide has already been produced and has escaped before the solution is consumed!) The antacid is actually the citrate ion, which reacts with stomach acid to produce citric acid. The citric acid is more easily consumed by the body, and because the citric acid holds the hydrogen ions more tightly than the stomach acid did, the overacidity of the stomach is relieved.

Rolaids®, another antacid, contain calcium carbonate. The carbonate ion reacts with two hydronium ions to produce carbonic acid and water. The carbonic acid then produces carbon dioxide and water:

$$CO_3^{2-} + 2\,H_3O^+ \rightarrow 2\,H_2O + H_2CO_3$$
$$H_2CO_3 \rightarrow H_2O + CO_2.$$

By reacting with hydronium ions in the stomach, the carbonate ions reduce the overacidity of the stomach. Stomach acid is a combination of gastric juices and hydrochloric acid.

WASTE AND THE ENVIRONMENT

The materials used in this investigation are not toxic. The acidic solutions can harm the plumbing if not diluted or neutralized.

▲ *CAUTION*s warn about safety hazards.
*EXTRA*s give helpful hints, additional information, or interesting facts.

Reagents
sodium bicarbonate [$NaHCO_3$]
0.5-M acetic acid [CH_3COOH]
0.5-M hydrochloric acid [HCl]

Common Materials
Alka-Seltzer®   baking powder
Rolaids®   baking soda

Laboratory Equipment
1000 mL beaker   ring stand, ring, and wire gauze
600 mL beaker   utility clamp
100 mL graduated cylinder   1-hole stopper to fit 100 mL graduated
10 mL graduated cylinder   cylinder
15 cm test tube   glass tube for stopper
rubber tubing (at least 24") that will fit snugly on the glass tubing

PROCEDURE

1. Set up the gas-measuring apparatus as shown below by filling a 1000 mL beaker with about 800 mL of water. Fill the 100 mL graduated cylinder with water. Place your fingers or palm tightly over the top and invert the cylinder into the beaker, letting as little air into the cylinder as possible. Then clamp the cylinder to a ring stand. Insert the rubber tubing up into the cylinder so that any gas passing through the tubing will go into the cylinder. Keep the stopper end elevated or the hose will act like a siphon allowing the water to drain out of the 1000 mL beaker through the hose.

2. Set a 600 mL beaker on an elevated ring on the ring stand. Place a dry 15-cm test tube in the beaker.

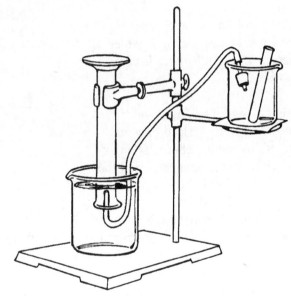

3. Weight out 0.34 g of sodium bicarbonate ($NaHCO_3$) onto a piece of weighing paper. Measure out 14 mL of 0.5-M acetic acid (CH_3COOH) in a graduated cylinder.

4. Read and record the water level in the cylinder. Record this value as the initial reading.

5. Pour the sodium bicarbonate into the test tube. Add the acid to the test tube, and place the stopper that is connected to the tubing in the test tube. Allow the reaction to continue until no more carbon dioxide is produced. Usually no more than 5 minutes is required.

⚠ *CAUTION*
Acetic acid is corrosive.

EXTRA
Zero is at the top.

EXTRA
Stopper quickly and tightly.

6. Read and record the water level in the cylinder. Record this reading as the final reading. Refill the graduated cylinder for the next trial.

EXTRA
Read zero at the top.

7. Repeat steps 3 through 6 using another 0.34 g of sodium bicarbonate.

8. Repeat steps 3 through 6 using 0.34 g of baking soda instead of sodium bicarbonate. Compare the amounts of CO_2 produced by sodium bicarbonate and by baking soda and use the greatest amount in future calculations.

9. Repeat steps 3 through 6 using about 0.34 g of an Alka-Seltzer tablet instead of sodium bicarbonate, and water instead of acetic acid. (0.34 g of an Alka-Seltzer is about one-fourth of a tablet.)

10. Repeat steps 3 through 6 using 0.5 M hydrochloric acid (HCl) instead of acetic acid, and about 0.34 g of a Rolaids tablet instead of sodium bicarbonate. (0.34 g of Rolaids is about one-third of a tablet.)

⚠ *CAUTION*
Hydrochloric acid is corrosive.

⚠ *CAUTION*
Don't burn yourself.

11. Repeat steps 3 through 6 using 0.34 g of baking powder instead of sodium bicarbonate, and water instead of acetic acid. Prepare a warm-water bath by putting about 300 mL of water in the 600 mL beaker. Measure the temperature of the water before adding the reactants. It should be at room temperature (about 20°C to 25°C). When the reactants are added, some CO_2 is produced. When carbon dioxide production ceases, record the water level in the cylinder as the cold water level. Using a Bunsen burner, heat the 600 mL beaker until the water temperature is about 90°C. Keep the water hot until the carbon dioxide production again ceases. Record the water level in the cylinder as the hot water level.

EXTRA
Isn't this lab a gas!

12. Flush the solutions down the drain with plenty of water.

PRODUCTION OF CARBON DIOXIDE
PRE-LAB QUESTIONS

Name: _____

Lab Partner: _____

Section: _____   Date: _____

1) If a recipe calls for lemon juice, would you expect it to require baking soda or baking powder?

2) If a recipe calls for both baking soda and baking powder, what is a likely reason?

3) Even after combining the ingredients, a certain recipe for bran muffins will keep for a week when refrigerated. As soon as the batter is placed in the oven, it rises. What ingredient provides the carbon dioxide gas?

4) Carbon dioxide gas is soluble in water. How will this affect the measurement of gas produced by each reaction during the investigation?

PRODUCTION OF CARBON DIOXIDE
REPORT SHEET

Name: _____ _____

Lab Partner: _____

Section: _____   Date: _____

I. SODIUM BICARBONATE ($NaHCO_3$)   #1   #2

    Sodium bicarbonate   _____ g   _____ g

    Final water level   _____ mL   _____ mL

    Minus initial water level   −_____ mL   _____ mL

    Volume of CO_2 produced   _____ mL   _____ mL

        Larger value $\dfrac{\text{mL } CO_2}{\text{g } NaHCO_3}$ = _____ mL/g

II. BAKING SODA

    Baking soda   _____ g

    Final water level   _____ mL

    Minus initial water level   −_____ mL

    Volume of CO_2 produced   _____ mL

    $\dfrac{\text{mL } CO_2}{\text{g baking soda}}$ = _____ mL/g

In the following calculations, use the largest of the three values from above for the mL of CO_2 produced by $NaHCO_3$. Baking soda is sodium bicarbonate ($NaHCO_3$).

III. BAKING POWDER

Baking powder   _____ g

Hot water level   _____ mL   Cold water level _____ mL

Minus initial water level   −_____ mL   Cold water temp._____ $^{\circ}$C

Volume of CO_2 produced   _____ mL   Hot water temp. _____ $^{\circ}$C

$$\frac{mL\ CO_2}{g\ baking\ powder} = \underline{\hspace{3cm}}\ mL/g$$

$$\frac{mL/g\ baking\ powder}{mL/g\ NAHCO_3} \times 100\% = \underline{\hspace{2cm}}\%\ NaHCO_3\ in\ baking\ powder$$

IV. ALKA-SELTZER

Alka-Seltzer   _____ g

Final water level   _____ mL

Minus initial water level   −_____ mL

Volume of CO_2 produced =   _____ mL

$$\frac{mL\ CO_2}{g\ Alka\text{-}Seltzer} = \underline{\hspace{3cm}}\ mL/g$$

$$\frac{mL/g\ Alka\text{-}Seltzer}{mL/g\ NAHCO_3} \times 100\% = \underline{\hspace{2cm}}\%\ NaHCO_3\ in\ Alka\text{-}Seltzer$$

V. ROLAIDS

Rolaids   _____ g

Final water level   _____ mL

Minus initial water level   −_____ mL

Volume of CO_2 produced =   _____ mL

$$\frac{mL\ CO_2}{g\ Rolaids} = \underline{\hspace{3cm}}\ mL/g$$

$$\frac{mL/g\ Rolaids}{mL/g\ NAHCO_3} \times 100\% \times \frac{100\ g\ CaCO_3}{84\ g\ NaHCO_3} = \underline{\hspace{3cm}}\%\ CaCO_3\ in\ Rolaids$$

VI. QUESTIONS

1. What would happen if you swallowed a whole Alka-Seltzer tablet?

2. What occurs if dough is left overnight before it is baked?

3. What is the difference between sodium bicarbonate from the laboratory and baking soda from the kitchen?

4. Would you ever use calcium carbonate ($CaCO_3$) as a substitute for sodium bicarbonate ($NaHCO_3$) in cooking? Why or why not?

5. If the percent $NaHCO_3$ determined in the investigation is greater than 100%, which of the following is a logical explanation?

   a. There is more than 100% $NaHCO_3$ in the material.

   b. Some of the CO_2 gas in the sodium bicarbonate determination was not captured.

   c. Extra gas was produced by a second reaction.

6. What happens if you are slow when inserting the stopper in the test tube or if the stopper does not make a good seal?

47 Paper Chromatography of Food Colors

What Color Is Your Food?

OBJECTIVES

To isolate the colors used in the outer coatings of hard candies and to identify the dyes used to create the various colors.

To use the method of paper chromatography to separate the colors into pure dyes.

To calculate the R_f values of pure dyes and of the isolated dyes.

To compare the R_f values of pure FD&C dyes with the separated colors from the candies.

Relates to Chapter 17 of Chemistry for Changing Times, *twelfth ed.*

BACKGROUND

Most of us have very definite ideas about what color our food is supposed to be. We like red cherries, orange oranges, and purple grape jelly, but candy can be many colors. The U.S. Food and Drug Administration controls the colors that can be used in our foods. At present, seven synthetic colors are allowed: FD&C blue #1 and #2, red #3 and #40, yellow #5 and #6, and green #3. Many of the dyes used in our foods are actually a combination of these dyes. The components of these colors can often be determined by chromatography.

One type of chromatography is *paper chromatography*, the method that will be used in this investigation. Paper chromatography works because of capillary action, the interaction of compounds with paper, and the solubility of compounds in solvents. It is easy to demonstrate capillary action. Place a paper towel across the edge of a beaker of water with the end of the towel in the water. The water will move up into the towel, rising up to the edge of the beaker and then down onto the table if given enough time.

Paper chromatography takes advantage of capillary action by allowing a solvent to rise along a piece of paper. The solvent will carry with it any compounds that are dissolved in it. The distance the compounds are carried depends on how soluble each compound is in the solvent and how much interaction or attraction each compound has for the paper. The more soluble the compound is in the solvent, the farther the compound travels. The more interaction with the paper, the less distance the compound travels.

The distance traveled by the compound is normally recorded by computing an R_f value. This value is a ratio of the distance traveled by the solvent to the distance traveled by the compound:

$$R_f = \frac{\text{Distance traveled by compound}}{\text{Distance traveled by solvent}}$$

When a dye is used in a food, sometimes the components of the food can interfere with the chromatography test. To ensure a correct determination, the dye needs to be separated from the other components. The separated dye will be taken up into a natural fiber. Although the procedure is written for wool, 100% cotton yarn will also work. Synthetic yarn will not work well as a different type of dye is required to color synthetic fibers than to color natural fibers.

The dye compounds are all fairly complex organic compounds. The oxygen, nitrogen, sulfite groups (SO_3^{2-}), hydroxides (OH^-), and ether groups (-O-) improve the water solubility of the compounds. Because of the differences among the dyes in water solubility and in the interaction of the compounds with the chromatography paper, the compounds have different R_f values.

FD&C Yellow #6
Sunset Yellow
CI 15985

FD&C Red #40
Allura Red AC
CI 16035

FD&C Yellow #5
Tartrazine
CI 19140

FD&C Red #3
Erythrosin B
CI 45430

FD&C Blue #1
Erioglaucine
CI 42090

FD&C Green #3
Fast Green
CI 42053

FD&C Blue #2
Indigo Carmine
CI 73015

In this investigation, you'll use a chromatography system to look at some common food dyes.

WASTE AND THE ENVIRONMENT
The compounds used in this investigation are not toxic. Acidic and basic solutions can harm the plumbing if not neutralized or diluted.

▲ *CAUTION*s warn about safety hazards.
*EXTRA*s give helpful hints, additional information, or interesting facts.

Reagents

butanol [$CH_3(CH_2)_3OH$]

ethanol [CH_3CH_2OH]

2-M ammonia [NH_3]

concentrated acetic acid [CH_3COOH]

1% ammonia solution [NH_3]

pH test paper

Common Materials

FD&C food colors: B1, B2, Y5, Y6, stapler
    R3, R40, G3

grocery store food colors (R, G, B, Y)

colored candies (M&M's® or Skittles®)

paper clips

stapler and staples

undyed wool yarn or fabric

stapler

Special Equipment

chromatography paper

plastic wrap or sealing wax (or Parafilm®)

10 mL graduated cylinder

150 mL beaker

watch glass

PROCEDURE

1. Prepare 100 mL of eluting solution by mixing 60 mL of butanol, 20 mL of ethanol, and 20 mL of 2-M ammonia (NH_3).

2. Obtain a piece of chromatography paper from your instructor. Be careful at all times to keep the paper clean and dry. Touch the paper only at the edges or with gloves as protection from the oils on your hands. With a pencil, make a line across the lower portion of the paper 1.5 cm from the bottom. Mark 7 *X*s on this line evenly spaced from one another but at least 2 cm from each outer edge. Label them B1, B2, Y5, Y6, R3, R40, G3 for the seven FD&C dyes. Mark a second line 7.5 cm from the bottom.

EXTRA
Ink from a ballpoint or felt tip may run during the experiment.

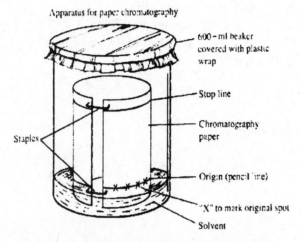

Apparatus for paper chromatography

600-ml beaker covered with plastic wrap

Stop line

Chromatography paper

Staples

Origin (pencil line)

"X" to mark original spot

Solvent

3. Straighten a paper clip so that you can use one end to stick into the dye and then to touch the corresponding *X*. A small drop of dye will adhere to the paper clip and then will be absorbed by the paper. Make each spot dark by applying dye to the same spot several times. Allow the spot to dry between applications. Clean the paper clip between dyes or you will contaminate the dye spots.

4. Cover the bottom of a 150 mL beaker with eluting solution to a depth of about 2 to 3 mm.

5. Roll the chromatography paper so the two edges just overlap. Staple or paper clip the top and the bottom to hold the edges together. The staples or paper clips should not touch any of the *X*s.

6. Gently place the paper in the solution. Do not allow the solution to touch the dots of dye or the paper to touch the walls of the beaker. Immediately cover the beaker with Parafilm®. Allow the solution to rise until the solvent front has traveled to the second line.

EXTRA
The container must be covered to prevent evaporation. The solvent can evaporate quickly enough to stop the chromatography.

7. Remove the paper, remove the staples or clips, and immediately mark the solvent front and outline the dye spots with pencil.

EXTRA
The next sample can be prepared while the solvent is rising in step 6.

8. Measure and record the center of each spot from the beginning line. Measure and record the distance of the solvent front from the beginning line. Calculate R_f values as outlined in the report sheet.

9. Repeat the procedure (steps 2 through 8) with another piece of paper, using as samples the four food colors (R, Y, G, B) that can be bought in the grocery store.

⚠ *CAUTION*
It is dangerous to eat in lab. Acetic acid is caustic.

10. Using one color in each test tube, prepare dyes from one or two colors of M & M's® or Skittles® or other brightly colored sugar-coated candy. Place 6 to 8 candies of one color in a test tube with 15 mL of distilled water. Add 5 drops of concentrated acetic acid.

CAUTION
Don't burn yourself. Don't point the test tube at anyone.

11. Heat very gently until the color coating dissolves. Dissolve as little of the inside white candy as possible.

12. Gently pour the colored solution into a beaker leaving all solids in the original test tube.

13. Add a 15-cm piece of un-dyed woolen yarn or a small (2-cm × 5-cm) piece of un-dyed wool fabric and 5 drops of concentrated acetic acid.

EXTRA
Wool in an acidic solution soaks up dye; 100% cotton also works.

14. Boil gently for 4 to 5 minutes or until all color is gone from the solution, stirring occasionally with a glass rod.

⚠ *CAUTION*
Don't burn yourself.

15. Rinse the wool well in tap water to remove all other components.

16. Place the wool in a small clean beaker. Add 5 mL of a 1% ammonia solution. Test the pH with pH paper. If the solution is not basic, add more ammonia solution.

EXTRA
Dye is released from the wool in a basic solution.

17. Boil for 4 to 5 minutes with occasional stirring or until the dye is in solution and the color of the yarn or fabric has faded significantly.

EXTRA
The color cannot be completely removed from the yarn.

18. Pour the solution onto a clean watch glass and evaporate to dryness over a water bath. During drying, occasionally run the stirring rod around the perimeter of the liquid to keep the dye in solution.

19. Allow the watch glass to cool. Add 2 drops of distilled water and stir with a stirring rod. This is the dye to use in the next step.

20. Repeat steps 2 through 8 using the dyes prepared from the candies.

21. Compare your results with those of other student pairs who are using other colors of the candies.

22. Throw the solids and yarn into the trash. Flush the solutions down the drain with plenty of water.

PAPER CHROMATOGRAPHY OF FOOD COLORS
PRE-LAB QUESTIONS

Name: _____

Lab Partner: _____

Section: _____   Date: _____

1) Why do you suppose the numbers of the FD&C-approved food dyes are not consecutive?

2) Would you expect stain removers to be acidic or basic?

3) Most synthetic fibers must be dyed during their production process. Why do natural fibers accept a dye, and synthetic fibers do not?

4) After cloth is dyed at home, it is often boiled in a vinegar solution. Why?

PAPER CHROMATOGRAPHY OF FOOD COLORS
REPORT SHEET

Name: _____

Lab Partner: _____

Section: _____ Date: _____

I. DYES Distance traveled (cm) $R_f = \dfrac{\text{Distance traveled by dye}}{\text{Distance traveled by solvent}}$

Solvent Distance: _____

B1 _____ _____

B2 _____ _____

Y5 _____ _____

Y6 _____ _____

R3 _____ _____

R40 _____ _____

G3 _____ _____

II. FOOD COLORS

	Distance traveled	R_f	Identity B1, B2, Y5, Y6, R3, R40, G3
Solvent distance:	_____		
Components of:			
Red	_____	____	_____
	_____	____	_____
Yellow	_____	____	_____
	_____	____	_____
Green	_____	____	_____
	_____	____	_____
Blue	_____	____	_____
	_____	____	_____

III. CANDY DYES

	Distance Traveled	R_f	Identity B1, B2, Y5, Y6, R3, R40, G3
Solvent distance:	_____		
Candy color			
1	_____	____	_____
2	_____	____	_____
3	_____	____	_____
4	_____	____	_____
5	_____	____	_____
6	_____	____	_____

IV. QUESTIONS

1. a. What is the difference between blue #1 and blue #2?

    b. Between red #3 and red #40?

    c. Between yellow #5 and yellow #6?

2. Are the food colors pure—that is, are they made of one dye?

3. Are any of the candy dyes made of only one dye?

4. Why does a colored acidic food make a worse stain on wool than a colored basic food?

48 *Detection of Alcohol*

I Had Only One Drink, Occifer

OBJECTIVES

To simulate "alcohol breath" by blowing through a solution of ethanol in water.

To make use of a control in testing the alcohol content of breath obtained from varying concentrations of ethanol.

To illustrate a quantitative measurement of the amount of alcohol present on the breath.

Relates to Chapters 9 and 18 of Chemistry for Changing Times, *twelfth ed.*

BACKGROUND

Ethyl alcohol is the most misused or "abused" drug. There are millions of people who feel that they must have something to drink. There are others who drink to relax, or to forget, or to have fun. Unfortunately, many of these people do not realize how much alcohol slows their reflexes, distorts their view, and clouds their reasoning.

One especially tragic outcome of alcohol use is *fetal alcohol syndrome*. A baby born to a mother who drinks while she is pregnant is often small, deformed, or mentally retarded.

It is especially dangerous for drinkers to attempt to drive an automobile. Alcohol is involved in about half of all fatal automobile accidents.

Alcohol enters the bloodstream directly through the lining of the stomach very soon after it is ingested. The blood then carries the alcohol to all organs of the body, including the lungs and the brain. Much of the alcohol is broken down in the liver, but the remainder is exhaled through the lungs or passed from the body in urine.

Today's instrumentation provides the means for police to test drivers for alcohol consumption. Before the advent of modern instrumentation, however, chemical reactions were used to prove intoxication. One such reaction that indicates the presence of alcohol uses the dichromate ion.

In this investigation, you will simulate alcohol content on the breath and then use a dichromate solution to detect the alcohol. This setup is not intended to be like a modern police alcohol detection system.

The reaction is:

$$3\ CH_3CH_2OH(g) + 2\ Cr_2O_7^{2-} + 16\ H^+ \longrightarrow 3\ CH_3CO_2H + 4\ Cr^{3+} + 11\ H_2O.$$

Yellow Orange   Green

The alcohol group (-OH) reacts with the dichromate ion ($Cr_2O_7^{2-}$) in an oxidation-reduction reaction. The alcohol is oxidized to a carboxylic acid (-COOH) while the dichromate is reduced to the chromium 3+ ion. Dichromate is a yellow orange color, whereas the solution of chromium (III) ions is green.

Dichromate is a very hazardous chemical. This experiment should be done only with great care.

WASTE AND THE ENVIRONMENT

Dichromate and the chromium ion are both classified as toxic chemicals. Sulfide is used in the disposal because the sulfides are very insoluble in water. Dichromate is also a very powerful oxidizing agent. Concentrated acidic or basic solutions can damage plumbing if not neutralized or diluted. The rest of the compounds are not toxic.

▲ *CAUTION*s warn about safety hazards.
*EXTRA*s give helpful hints, additional information, or interesting facts.

Reagents

potassium dichromate [$K_2Cr_2O_7$]
sodium carbonate [Na_2CO_3]
ethanol (ethyl alcohol) [CH_3CH_2OH]
acetone [CH_3COCH_3]
sodium sulfide [Na_2S]
thioacetamide [CH_3CSNH]

concentrated (18 M) sulfuric acid [H_2SO_4]
3-M sodium hydroxide [NaOH]
1-M ferric chloride (iron(III) chloride)
   [$FeCl_3$]
pH test paper

Laboratory Equipment

various lengths of glass tubing and rubber hose
500 mL Erlenmeyer flask
2-hole stopper to fit 500 mL flask
funnel

filter paper
125 mL Erlenmeyer flask
600 mL beaker

PROCEDURE

1. Place 300 mL of water into a 600 mL beaker. Add 30 g of potassium dichromate ($K_2Cr_2O_7$) and 30 mL of concentrated (18-M) sulfuric acid (H_2SO_4). Swirl until the solid is dissolved. Label the beaker "dichromate solution" and set it aside.

▲ *CAUTION*
$K_2Cr_2O_7$ is a hazardous chemical. Sulfuric acid is caustic.

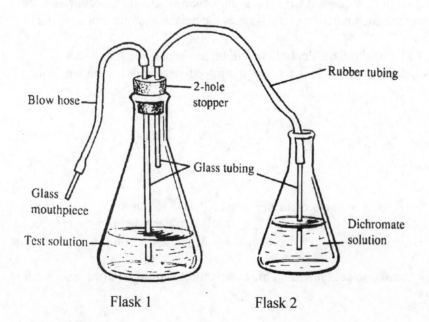

Blow hose

2-hole stopper

Rubber tubing

Glass tubing

Glass mouthpiece

Test solution

Dichromate solution

Flask 1   Flask 2

2. Arrange a 500 mL Erlenmeyer flask with stopper and hoses as shown in the drawing. Be sure to wash the mouthpiece well. It is better to use disposable straws in place of the glass mouthpiece.

3. Prepare a control sample by adding 0.1 mL of ethanol to 50 mL of dichromate solution. Set the control sample aside.

4. Place 50 mL of the dichromate solution in a 125 mL Erlenmeyer flask. Place the glass end of the test hose in the dichromate solution (flask 2).

5. Place 100 mL of test solution in the 500 mL flask (flask 1). The glass tube from the blow hose should be immersed in the solution. The first test solution should be 50 mL of ethyl alcohol and 50 mL of water.

6. One lab partner should time 60 seconds on a watch as the other partner blows through the mouthpiece with long, even breaths.

⚠ *CAUTION*
Do not suck the solution into your mouth. Remove the mouthpiece to inhale.

7. As soon as the student stops blowing, begin timing again. Set the flask beside the control sample. Stop timing when the test sample is the same color as the control sample.

8. Repeat steps 4 through 7 for a 40 mL alcohol, 60 mL water solution.

9. Repeat steps 4 through 7 for a 30 mL alcohol, 70 mL water solution.

10. Repeat steps 4 through 7 for a 20 mL alcohol, 80 mL water solution.

11. Repeat steps 4 through 7 for a 10 mL alcohol, 90 mL water solution.

12. One student may clean up while the other times the last test solution. It may not change color before lab time is finished.

13. Add alcohol to any excess dichromate solution. The dark green chromium solution must be treated as waste.

14. Precipitate the chromium with a threefold excess of sodium sulfide (Na_2S) or thioacetamide (CH_3CSNH_2). You should began with 30 g of potassium dichromate ($K_2Cr_2O_7$), which

represents 0.20 mol of Cr. That means that complete precipitation would be ensured with 47 g of Na_2S in solution or 45 g of thioacetamide.

15. Let the solution sit for an hour. Check the pH. If it is acidic, adjust it to neutral with 3 M sodium hydroxide. Both of these steps are to ensure complete precipitation. Filter and discard the precipitate and filter paper in the chromium waste container.

16. The sulfide is hazardous and must be removed. To the filtrate, add 1-M ferric chloride (iron (III) chloride, $FeCl_3$) until no more solid forms.

17. Test the pH. If the solution is acidic, add sodium carbonate (Na_2CO_3) until it is neutral.

18. Filter the solution by pouring it through filter paper prepared as shown in the figures. Place the precipitate and the filter paper in the waste container marked iron (III) sulfide to be buried in a hazardous-waste landfill. Flush the remaining solution down the drain with plenty of water.

⚠ CAUTION
Sodium hydroxide is caustic.

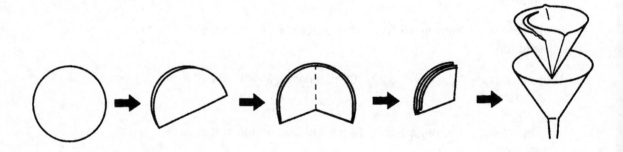

DETECTION OF ALCOHOL
PRE-LAB QUESTIONS

Name: _____

Lab Partner: _____

Section: _____   Date: _____

1) How could blowing through an alcohol solution simulate the alcohol content of a person's breath?

2) How would you expect the varied concentrations of alcohol to effect the time required for the reduction of the dichromate ion?

3) Why is a chemically based test not as reliable as an electronically based alcohol test?

4) Discuss ways in which the results of a test of the sort used in this investigation might vary between individuals.

DETECTION OF ALCOHOL
REPORT SHEET

Name: _____

Lab Partner: _____

Section: _____   Date: _____

I.   Test Solution

Time to Reach Color
of Control

50 mL alcohol, 50 mL water   _____ min

40 mL alcohol, 60 mL water   _____ min

30 mL alcohol, 70 mL water   _____ min

20 mL Alcohol, 80 mL water   _____ min

10 mL alcohol, 90 mL water   _____ min

II. QUESTIONS

1. What is the color of the control sample?

2. What happens to the temperature of the dichromate solution? Why?

3. Would this test be convenient for a police officer to carry if it were calibrated to determine percent alcohol?

4. Why was proper disposal of the dichromate and chromium solution important?

5. Why would a chromium solid that is more soluble be less satisfactory for disposal?

49 *Aspirin Preparation and Purity Test*

This Lab Gives Me a Headache!

OBJECTIVES

To prepare and purify aspirin from acetic anhydride and salicylic acid.

To test commercial aspirin and the prepared aspirin for free salicylic acid by thin-layer chromatography.

To calculate a percent yield for the aspirin product.

To understand that phase changes such as recrystallization, can be used as a purification process.

To appreciate the usefulness of an R_f value in analytical applications.

Related to Chapters 9, 18, and 19 of Chemistry for Changing Times, *twelfth ed.*

BACKGROUND

Aspirin is the most common non-prescription pain reliever. Its systemic name is acetylsalicylic acid. It is a white solid that is almost insoluble in cold water. Salicylic acid is also an *analgesic* (a pain reliever), but it is irritating to the stomach. Changing the OH group on the benzene ring to an ester group, as shown in the following reaction, can reduce the irritation. Because this group has two carbons, it is called an *acetyl group*. Several acetylating agents can achieve the reaction but acetic anhydride is inexpensive and not too difficult to handle in the laboratory. The reaction is:

| Salicylic Acid | Acetic Anhydride | Acetylsalicylic Acid "Aspirin" | Acetic Acid |

Sulfuric acid is used as a *catalyst*. A catalyst causes the reaction to occur more rapidly but without being consumed. One of the reaction products is aspirin and it can be purified by crystallization and filtration. The aspirin product is less soluble in cold water than are the starting materials or other products. These impurities can then be separated by filtration and disposed of with the filtrate. The aspirin can then be dissolved and recrystallized to further purify it. The change of state is a purification process because as a substance recrystallizes, it tends to exclude impurities if the experimental conditions are adjusted correctly.

A breakdown of aspirin will occur with time, generating salicylic acid and acetic acid. The degenerative process is accelerated in the presence of moisture. Since salicylic acid is irritating, it is not supposed to be present in a commercial aspirin product. An old bottle of aspirin will sometimes have a faint smell of vinegar. Vinegar is dilute acetic acid. A more accurate way of testing for aspirin deterioration is to test for the presence of salicylic acid by liquid chromatography.

Liquid chromatography is a separation technique using a solvent solution that will climb up the surface of *silica gel*. The silica gel is applied in a very thin layer to a sheet of clear acetate and the production of a chromatogram on a surface of this type is referred to as a *thin-layer chromatography*, or TLC. If the silica has absorbed a spot of a chemical mixture, each of the components will be carried along with the solvent to a certain extent or distance. Each component will appear as a spot on the TLC plate some distance up the plate from the point of sample application. The distance each component is carried by the solvent depends upon the interactions between the compound and the solvent, the interactions between the compound and the silica, and the molecular weight of the compound. We can quantify this information by the equation

$$R_f = \frac{\text{distance of spot center from baseline}}{\text{distance solvent traveled from baseline}}$$

where R_f is the relative distance traveled by a substance, and the baseline is the initial location of the spot. R_f values are common analytical tools for identifying components of a mixture.

WASTE AND THE ENVIRONMENT

The eluting solution is toxic, so it should not be placed into the water system. However, it is volatile and can be evaporated in a hood. The prepared aspirin and the solutions used to prepare the aspirin are not toxic.

⚠ *CAUTION*s warn about safety hazards.

*EXTRA*s give helpful hints, additional information, or interesting facts.

Reagents

salicylic acid [$C_7H_6O_3$]	concentrated sulfuric acid [H_2SO_4]
acetic anhydride [$(CH_3CO)_2O$]	95% ethyl alcohol [CH_3CH_2OH]
eluting solvent (66:33:1 hexane, ethyl acetate, glacial acetic acid)	

Common Materials

fresh commercial aspirin	ice
aged aspirin	paper towels, plastic wrap, or foil (to fit over beakers)

Laboratory Equipment

3 paper clips	hair dryer
TLC chromatography plates	filter paper, large size for drying
ultraviolet light	Buchner funnel
wash bottle of distilled water	filter paper for Buchner funnel
2 600 mL beakers	balance
125 or 250 mL Erlenmeyer flask	stirring rod
150 mL beaker	4 small test tubes
250 mL beaker	mortar and pestle
hot plate	thermometer

PROCEDURE

I. Preparation of Aspirin

1. A team of four students is suggested for this investigation. Prepare two water baths. Make a hot-water bath by heating about 400 mL of water to near boiling in a 600 mL beaker. Make an ice bath in a second 600 mL beaker. Place a wash bottle of distilled water in the bath to cool.

2. While the water is heating, accurately weigh about 9 g of salicylic acid and record the mass to the nearest 0.01 g.

3. Measure and pour 12 mL of acetic anhydride into a 125 or 250 mL Erlenmeyer flask. Add the salicylic acid and blend by swirling.

4. Add the catalyst, 12 to 14 drops of concentrated sulfuric acid. Swirl again and suspend the flask in the hot-water bath (a clamp is necessary to prevent the water from getting into the reaction flask). Bring the bath to a gentle boil and allow the reaction to proceed for 12 to 16 minutes. Shake or stir occasionally while heating. The aspirin formed should be in solution at this point.

5. Remove the flask from the boiling water and cool it under running tap water over the outside of the flask until the solution is at or below room temperature. Avoid dilution of the flask contents with water while cooling.

6. Add 30 mL of cold distilled water (below 4°C) to the reaction mixture and set the flask in a beaker of ice to encourage complete crystallization.

7. Separate the crystals from the liquid by suction filtration, using the Buchner funnel suction apparatus as shown in the sketch. Wet the filter paper with distilled water to seat it before adding the mixture.

EXTRA
It will be handy to keep the wash bottle chilling in the ice bath.

⚠ *CAUTION*
Acetic anhydride is irritating to skin and eyes.

⚠ *CAUTION*
Don't burn yourself. Sulfuric acid is very caustic.

EXTRA
Too much sulfuric acid will result in a dense yellow oily substance and you will need to start over.

EXTRA
Be sure to use chilled water to filter.

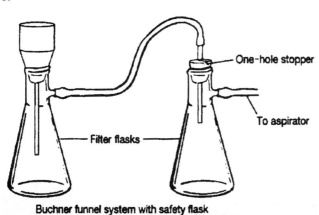

Buchner funnel system with safety flask

8. Draw air through the mat of crystals on the filter paper to partially dry the crystals. Transfer the crystals to a 150 mL beaker. Discard the filtrate.

9. Add 30 mL of 95% ethanol to the crystals, then, if necessary, warm the beaker (in a water bath or on a hot plate) to dissolve the crystals. **Do not boil.** If the crystals still do not dissolve, consult the instructor. Then add up to 5 mL more ethanol in 1-mL increments, waiting 4 to 5 minutes between additions.

⚠ CAUTION
Don't burn yourself. Ethanol is flammable.

EXTRA
Keep the ethanol volume to a minimum. This will encourage more complete crystallization later.

10. If any undissolved solids remain, these may be impurities. Set up a clean filtering system as in step 7. Filter the ethanol/aspirin solution to remove the solid impurities.

11. Pour 50 to 60 mL of warm water into the ethanol solution, cover the beaker with a watch glass, and set the solution aside to cool and crystallize.

⚠ CAUTION
The aspirin you prepare may not be free of contaminants and should not be ingested.

12. Complete the cooling in a beaker of ice and water. Filter again using chilled water. The air-dried crystals from the filter paper can be dried more completely by pressing them between two sheets of a larger size filter paper than what was used in the Buchner funnel.

13. When the crystals are completely dry, scrape them into a pre-weighed 100 mL beaker and weigh. Calculate the percent yield, and after testing for free salicylic acid, turn in your product.

II. Analysis of Aspirin for Free Salicylic Acid

1. Prepare your chromatography developing chamber by pouring about 5 mL of the eluting solvent into a 250 mL beaker. The eluting solvent is an approximately 66:33:1 mixture of hexane, ethyl acetate, and glacial acetic acid. Cover the beaker with a paper towel or aluminum foil to prevent the rapid evaporation of the eluting fluid.

2. Obtain a piece of TLC plate. Do not touch the plate directly. The compounds on your hands will interfere with the investigation. Handle it with plastic wrap or paper. Place it on a clean surface such as a paper towel. Measure along the longer sides of the plate and mark the plate with a pencil at 1.5 cm from the lower end and 1 cm from the top end. Connect the marks across the plate. The bottom (1.5-cm) line is the baseline, and the top line is the desired level for the solvent to reach. Place four evenly spaced *X*s on the bottom line, starting at least 1 cm from the side.

CAUTION
The solvent is toxic and flammable. Avoid breathing the vapors. Use a hood if available.

EXTRA
The chromatography plate can be cut to 2" x 4". Do not press hard with the pencil or the gel will flake off of the plate.

Apparatus for Chromatography

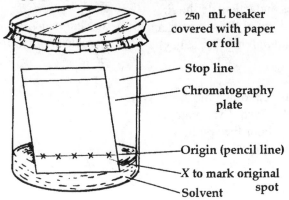

- 250 mL beaker covered with paper or foil
- Stop line
- Chromatography plate
- Origin (pencil line)
- X to mark original spot
- Solvent

3. Prepare 4 spotting sample tubes in 4 small test tubes by placing 4 to 5 mL of ethanol in each. To one tube, add a pinch of solid salicylic acid; to the second, add half of a fresh crushed commercial aspirin tablet; and to the third, add half of an aged commercial aspirin tablet that has been crushed. To the fourth tube, add a pinch of the aspirin prepared in Part I. Stopper the test tubes and mix them for 3 minutes to dissolve all the substance that will dissolve. Allow the undissolved substances to settle.

EXTRA
The aged tablet should be more than a year old, or tablets may be aged in a closed container at 100°C for a few days.

4. Using a clean, straightened paper clip for each sample as a spotting tool, spot each X on the paper with the corresponding solution. Try to keep the spots less than 3 mm in diameter. Allow the spots to dry, and repeat the application 6 to 8 times to build up a concentration on the spots. A hair dryer may be used to speed up the drying process.

5. Place the plate in the developing chamber and cover with paper or foil to prevent evaporation of the eluting fluid.

6. Allow the solvent to rise until it reaches the top line or your instructor tells you to stop. It should develop in less than 15 minutes.

7. Remove the plate from the solvent and allow it to dry or dry it with a hair dryer in the hood. Observe the plate under UV light. A shortwave UV light will cause the salicylic acid to appear light blue to white and the aspirin brown. Mark the distance the solvent traveled and the distance of each spot. Measure the distance to the center of each spot. Record the distances and calculate the R_f values. The spot for salicylic acid should travel the same distance (have the same R_f value) in any sample containing free salicylic acid.

8. Flush the solutions used to prepare the aspirin down the drain with plenty of water. Throw the aspirin into the trash. Evaporate the eluting solution in an almost closed hood. Pour the spotting solutions down the drain with plenty of water.

NOTE
Do not leave the paper clip in the salicylic acid solution because it attacks the metal and gives a purple tint to the solvent. The color does not affect results, however.

EXTRA
A white powder should be visible at each spot.

EXTRA
If uncovered, the solvent evaporates too quickly to climb the plate.

ASPIRIN PREPARATION AND PURITY TEST
PRE-LAB QUESTIONS

Name: _____

Lab Partner: _____

Section: _____   Date: _____

1) Why should you discard aspirin once its expiration date has been reached or if it has been stored in a moist or warm environment?

2) If a solid impurity fell into the beaker during the initial formation of the aspirin, in which step would it be removed?

3) If two separate samples of the same compound are analyzed by chromatography, why should the R_f values match even if the solvent fronts move different distances on the two chromatograms?

4) Why is ethanol used to dissolve and spot the samples on the chromatogram?

ASPIRIN PREPARATION AND PURITY TEST
REPORT SHEET

Name: _____

Lab Partner: _____

Section: _____   Date: _____

I. Mass of paper + salicylic acid   _____ g

    Minus mass of paper   − _____ g

    Mass of salicylic acid   _____ g

II. Mass of 100 mL beaker + aspirin   _____ g

    Minus mass of 100 mL beaker   − _____ g

    Mass of aspirin   _____ g

III. % Yield = $\dfrac{\text{experimental mass of product (aspirin)}}{\text{theoretical mass of product (aspirin)}} \times 100\%$

    % Yield = $\dfrac{\underline{\hspace{2cm}}\text{ g aspirin (II)}}{\underline{\hspace{2cm}}\text{ g salicylic acid (I)}} \times \dfrac{138.12\,\text{g/mol}}{180.15\,\text{g/mol}} \times 100\% = \underline{\hspace{2cm}}\%$

Instructor's initials for aspirin _____

IV. Describe the appearance of the salicylic acid spot.

Based on the TLC data, comment on the presence or absence of salicylic acid in the following:

1. Fresh commercial aspirin

2. Aged commercial aspirin

3. The aspirin prepared in this investigation

		Fresh	Aged	Today's Preparation
Salicylic Acid Distance traveled	Spot	_____	_____	_____
	Solvent	_____	_____	_____
	R_f	_____	_____	_____
Aspirin Distance traveled	Spot	_____	_____	_____
	Solvent	_____	_____	_____
	R_f	_____	_____	_____

V. QUESTIONS

1. Why is the yield not 100%?

2. Was excess water still in the aspirin when you weighed it?

3. Could another compound that is also insoluble in cold water be separated from aspirin by recrystallization?

4. Does the aspirin you produced smell of vinegar?

5. Should the aspirin you produced smell of vinegar?

6. Why does aspirin that has a sour smell have a greater chance of causing an upset stomach?

50 *Fats and Iodine Number*

To Saturate or To Unsaturate

OBJECTIVES

To detect the presence of fat in several foods by the oxidation of glycerol to acrolein.
To obtain a quantitative determination of the amount of fat in a food using iodine number.

Relates to Chapters 16, 17, and 19 of Chemistry for Changing Times, *twelfth ed.*

BACKGROUND

Fats are an important part of the human diet. The body uses fat as padding to protect the vital organs, such as the heart and liver. Fat is also used as insulation to prevent the body from changing temperature too quickly. The body also uses fat for energy storage. Although a small amount of fat is needed for a healthy diet, most Americans eat too much fat. This has led to a nation of overweight people and the health problems brought on by obesity. As an example of how too much fat can become a problem, the padding around the heart can cause pressure on the heart if the fat layer grows too thick. Excess fat in a diet can also contribute to blood circulation problems.

Medical science has discovered that certain fats are more healthy to consume than others. Fats with double bonds between carbon molecules are called *unsaturated*. Unsaturated fats are found in most vegetable oils and are healthier to consume. The saturated fats, which usually come from animal sources, have been linked with clogging of arteries and a higher risk of a heart attack. Many people are changing their diet to reduce their intake of saturated fat. Vegetable oils that are normally liquid can be *hydrogenated* to cause them to be solids at room temperature. During the process of hydrogenation, some or most of the multiple bonds are broken and hydrogen is added to fill the bonding sites on the carbon chains forming unsaturated fats. There is evidence that these "transformed" fats, called *trans fats*, may pose their own health problems.

A measurement of the amount of unsaturation is the *iodine number*. The iodine number is the number of grams of iodine that reacts with 100 g of the unsaturated fat. The more double bonds in the fat, the larger the iodine number. Each iodine molecule reacts by breaking one of the double bonds and adding to the molecule.

$$-\overset{\overset{\displaystyle H}{|}}{\underset{\underset{\displaystyle H}{|}}{C}}-\overset{\displaystyle H}{C}=\overset{\displaystyle H}{C}-\overset{\overset{\displaystyle H}{|}}{\underset{\underset{\displaystyle H}{|}}{C}}- \quad + \quad I_2 \quad \longrightarrow \quad -\overset{\overset{\displaystyle H}{|}}{\underset{\underset{\displaystyle H}{|}}{C}}-\overset{\overset{\displaystyle H}{|}}{\underset{\underset{\displaystyle I}{|}}{C}}-\overset{\overset{\displaystyle H}{|}}{\underset{\underset{\displaystyle I}{|}}{C}}-\overset{\overset{\displaystyle H}{|}}{\underset{\underset{\displaystyle H}{|}}{C}}-$$

Violet   Colorless

A method of identifying fats is the production of acrolein by oxidation of glycerol. A fat is hydrolyzed by strong acid to a fatty acid and glycerol. The glycerol is dehydrated by potassium hydrogen sulfate to acrolein, which has a sharp irritating odor and is a severe pulmonary and skin irritant. It can sometimes be noticed when fatty meats are grilled—for example, in a poorly ventilated hamburger stand. The equations associated with this test are given below.

$$\begin{array}{c} \text{H}\quad\text{O}\\ |\quad\ \ ||\\ \text{H-C-O-C-}(\text{CH}_2)_n\text{CH}_3\\ |\quad\ \ \text{O}\\ |\quad\ \ ||\\ \text{H-C-O-C-}(\text{CH}_2)_n\text{CH}_3 \quad + \quad \text{H}_2\text{O} \quad\xrightarrow{\ \text{H}^+\ }\quad \overset{\text{O}}{\overset{||}{\text{H-O-C-}(\text{CH}_2)_n\text{CH}_3}} \quad + \\ |\quad\ \ \text{O}\\ |\quad\ \ ||\\ \text{H-C-O-C-}(\text{CH}_2)_n\text{CH}_3\\ |\\ \text{H} \end{array}$$

Fat

Fatty Acid

$$\begin{array}{c}\text{H}\\|\\ \text{H-C-O-H}\\ |\\ \text{H-C-O-H}\\ |\\ \text{H-C-O-H}\\ |\\ \text{H}\end{array}$$

Glycerol

$$\begin{array}{c}\text{H}\\|\\ \text{H-C-O-H}\\ |\\ \text{H-C-O-H}\\ |\\ \text{H-C-O-H}\\ |\\ \text{H}\end{array} \quad\xrightarrow{\ \text{KHSO}_4\ }\quad \begin{array}{c}\text{H-C=O}\\ |\\ \text{H-C}\\ ||\\ \text{H-C}\\ |\\ \text{H}\end{array}\quad +\quad 2\,\text{H}_2\text{O}$$

Glycerol

Acrolein

WASTE AND THE ENVIRONMENT

The hexane and iodine used in this investigation are classified as toxic substances. Iodine is a strong oxidizing agent. When used as an antiseptic, it kills germs by oxidation, similar to the way chlorine sanitizes swimming pools. Hexane is volatile.

▲ *CAUTION*s warn about safety hazards.

*EXTRA*s give helpful hints, additional information, or interesting facts.

Reagents

glycerol

potassium hydrogen sulfate [KHSO$_4$]

hexane [C$_6$H$_{14}$]

5% iodine solution [I$_2$ in C$_6$H$_{14}$]

concentrated sulfuric acid [H$_2$SO$_4$]

50% sodium thiosulfate [Na$_2$S$_2$O$_3$]

3-M sulfuric acid [H$_2$SO$_4$]

sodium hydroxide solution [NaOH]

pH test paper

Common Materials

vegetable oil

solid margarine

liquid margarine

peanut or olive oil

potato chip

lean meat

radish

Fritos®

butter

Laboratory Equipment

small test tubes

stirring rod

watch glass

PROCEDURE
Part I: Test for Fat

1. Place 3 drops of glycerol in a dry, clean test tube, 3 drops of vegetable oil in a second tube, and 3 drops of liquid margarine in a third tube.

2. Add several crystals of potassium hydrogen sulfate ($KHSO_4$) to each test tube. Heat the test tubes over a flame until the color just begins to change. If a hood is available, use it.

3. Carefully waft, or fan, the fumes from each test tube toward your nose. Record your observations.

4. On a watch glass, test a potato chip, a slice of potato, a slice of lean meat, a slice of radish, and Fritos for fat by adding 10 drops of concentrated sulfuric acid (H_2SO_4) to a 5-mm-square slice of each food. A yellow color is a positive test for fat.

5. Repeat step 4 for any other type of food your instructor makes available.

Part II: Determination of Relative Unsaturation

6. Add 3 mL of the liquid margarine to a test tube containing 5 mL of hexane. Mix the solution with a clean stirring rod for several minutes.

7. Add a drop of 5% iodine solution, stir for a few seconds. If the solution is clear, continue to add iodine 1 drop at a time with stirring. When the color remains, record the number of drops of iodine added to that fat.

8. Repeat steps 6 and 7 for butter, solid margarine, vegetable oil, and peanut or olive oil.

9. Run a blank by adding nothing to the hexane and repeating step 7.

10. Place all foods in the trash. The hexane may be evaporated in a hood. The iodine has to be reduced with a twofold excess of 50% sodium thiosulfate solution. Add 3-M sulfuric acid (H_2SO_4) until the pH is between 2 and 3. Set it aside for an hour. Neutralize the solution with sodium hydroxide (NaOH). Flush it down the drain with plenty of water.

⚠ *CAUTION*
Don't burn yourself. Don't point the test tube at anyone.

⚠ *CAUTION*
Don't sniff directly from the tube.

⚠ *CAUTION*
Sulfuric acid is corrosive.

⚠ *CAUTION*
Hexane is flammable. Make sure there are no open flames in the lab.

EXTRA
5% iodine is made by adding 5 g of I_2 to 95 mL of hexane.

EXTRA
This step shows that the hexane does not add to the iodine number.

EXTRA
Sodium thiosulfate is photographer's "hypo."

FATS AND IODINE NUMBER
PRE-LAB QUESTIONS

Name: _____

Lab Partner: _____

Section: _____   Date: _____

1) Industry can adjust whether a material is solid or liquid by adding or deleting double bonds. Which is done to produce a solid?

2) Should a person be worried about the type of fat found in solid vegetable margarine? Explain.

3) Many Mediterranean cultures dip their bread in olive oil mixed with roasted garlic and other spices. Compare the health value of this practice to the traditional bread and butter (or margarine) we use in the U.S.

4) Would you expect the iodine number to be greater for squeezable margarine or for tub margarine? Why?

FATS AND IODINE NUMBER
REPORT SHEET

Name: _____

Lab Partner: _____

Section: _____   Date: _____

I. FAT TEST
  1. Record your observations of each test tube.
     Test tube 1

     Test tube 2

     Test tube 3

  2. Record your observations of each food tested for fat.
     Potato chip

     Raw potato

     Radish

     Meat

     Fritos®

II. UNSATURATION

	# of Drops of I_2
Liquid margarine	_____
Butter	_____
Solid margarine	_____
Vegetable oil	_____
Peanut oil	_____

III. QUESTIONS

1. Rank the following in terms of increasing order of dietary health: butter, solid margarine, liquid margarine.

2. Why are so many products advertised as being fried in peanut oil or olive oil?

3. Where have you smelled burned fat before?

4. Would you expect a positive test for fat on a piece of celery?

51 *Carbohydrates*

Sugars and Starch

OBJECTIVES

To understand the divisions and structures of basic carbohydrates.

To confirm the presence of carbohydrate in several foods.

To determine the type of carbohydrate in a sample based on a series of tests, including Benedict's, Barfoed's, and iodine.

To demonstrate the action of salivary amylase on carbohydrates.

Relates to Chapters 16, 17, and 19 of Chemistry for Changing Times, *twelfth ed.*

BACKGROUND

Carbohydrates are the body's source of quick energy. Carbohydrates come in several forms, but the final outcome is the same for all of them. The body breaks all carbohydrates down into glucose units that are then used as fuel for the body.

All carbohydrates are made up of *monosaccharides*. Monosaccharides are classified by the number of carbons they contain. For example, the class *hexose* includes molecules with six carbon atoms and all have the basic formula $C_6H_{12}O_6$. Common examples are: glucose, fructose, and galactose. Monosaccharides can exist in two forms: open-chains and rings.

Glucose

Fructose

Carbohydrates are divided into two categories: simple sugars and complex carbohydrates. The *simple sugars* include monosaccharides and disaccharides. *Disaccharides* consist of two monosaccharides bonded together. *Sucrose* (table sugar) is made of a glucose unit bonded to a fructose unit. *Maltose* (two glucose units) and *lactose* (glucose and galactose) are also members of the disaccharide group. The enzymes in the digestive system can usually break the bond between the monosaccharide units within disaccharides. *Lactose intolerance* is a deficiency of the enzyme lactase that cleaves lactose into its two monosaccharides. The bond between monosaccharides in glucose and galactose can also be cleaved by the reaction of an acid, or by heat.

Complex carbohydrates are polymers of monosaccharide units in the ring form. *Amylose* and *amylopectin* are the two polysaccharides that make up starch. Both molecules are polymers of glucose. These polysaccharides can also be broken down to monosaccharides by acids, heat, and the human digestive system. Another example of a complex carbohydrate is *cellulose*. Although cellulose is also a polymer of glucose units, the bonding is slightly different. This slight difference is enough to prevent human digestion from breaking the bonds, although some animals and insects such as termites can digest cellulose.

Some nutrition experts believe that we would be healthier if our diet contained a smaller percentage of simple sugars and a greater percentage of complex carbohydrates.

Benedict's test uses Cu^{2+} in a basic solution to detect monosaccharides and the disaccharides that contain aldehyde groups in their open forms. Glucose has an aldehyde group in its open form. A positive test is the appearance of an orange color from the presence of Cu_2O. The reaction is

$$2\ Cu^{2+}\ +\ \underset{\text{Aldehyde}}{-\overset{\overset{\displaystyle O}{\|}}{C}-H}\ +\ 2\ H_2O\ \longrightarrow\ Cu_2O\ +\ \underset{\text{Carboxylic Acid}}{-\overset{\overset{\displaystyle O}{\|}}{C}-O-H}\ +\ 4\ H^+$$

Barfoed's test uses the copper 2+ ion in an acetic acid solution. Unlike Benedict's reagent, which reacts with monosaccharides or disaccharides in the same length of time, Barfoed's reagent reacts with monosaccharides more quickly than with disaccharides. This reagent can be used to distinguish between monosaccharides and disaccharides.

The iodine test produces a dark blue solution when starch is present. The amylose helix structure is just the right size to hold an iodine molecule in the right position to cause the dark blue color.

In this investigation, you will test several foods to determine the type of carbohydrate they contain.

WASTE AND THE ENVIRONMENT

Copper ion is a metal ion that should not be allowed into the water system. Iodine is a strong oxidizing agent. The other substances are not toxic.

⚠ *CAUTION*s warn about safety hazards.
*EXTRA*s give helpful hints, additional information, or interesting facts.

Reagents
Benedict's reagent
Barfoed's reagent
1% glucose solution
1% fructose solution
1% sucrose solution
1% cellulose
pH test paper

1% starch solution
iodine solution
pH 7 buffer solution
50% sodium thiosulfate solution [$Na_2S_2O_3$]
3 M sulfuric acid [H_2SO_4]
sodium hydroxide solution [NaOH]

Common Materials
1% table sugar, solution
1% honey, solution

1% powdered sugar, solution
1% corn syrup, solution

Laboratory Equipment
5 large test tubes

50 mL beaker

PROCEDURE

Part I: Tests for Simple Sugars

1. Place 8 drops of Benedict's reagent into each of 5 test tubes labeled "glucose," "fructose," "sucrose," "starch," and "cellulose." Place 5 mL of each of the corresponding 1% solutions in each test tube. Set up a boiling water bath. Place the tubes in the boiling water bath for 3 minutes. Remove and cool the test tubes. Observe and record the colors.

2. Place 2 mL of each of the 1% solutions of glucose, fructose, sucrose, starch, and cellulose in labeled test tubes. Add 2 mL of Barfoed's reagent to each test tube. Place the test tubes in a boiling-water bath. Record the time. Watch each tube for a color change. Remove each tube as a color change occurs. Remove all test tubes after 15 minutes. Record the time and color change.

3. Place 2 mL of each of the same 1% solutions as used in steps 1 and 2 into labeled test tubes. Add a few drops of aqueous iodine solution to each to test for starch.

4. Repeat steps 1 through 3 using 1% solutions of table sugar, honey, corn syrup, and powdered sugar.

Part II: Enzyme Action on Carbohydrates

5. Each person should collect about 1 mL of saliva in a clean test tube. Pour the saliva from both partners into a small beaker. A commercial amylase solution may be used in place of the saliva. Rinse both test tubes into the beaker with about 10 mL of pH 7 buffer solution. Add enough distilled water to bring the total volume to 30 mL of solution in the beaker. Mix thoroughly.

6. Pour 3 mL of the saliva solution into each of 4 large test tubes.

7. To test tube 1, add 10 mL of 1% starch solution. To test tube 2, add 10 mL of 1% sucrose solution. To test tube 3, add 10 mL of 1% table sugar solution. To test tube 4, add 10 mL of 1% cellulose solution. Mix the contents of each test tube thoroughly. Set the test tubes aside for 15 minutes. Add 8 drops of Benedict's reagent to each. Place the test tubes in the boiling-water bath. The reaction may be slower than in step 1.

⚠ _CAUTION_
Don't burn yourself.

EXTRA
Monosaccharides change color in 2 to 3 minutes. Disaccharides change in about 10 minutes.

EXTRA
Think of pickles or your favorite foods to stimulate saliva production.

EXTRA
An enzyme in saliva breaks down starch and disaccharides into simpler sugars.

419

8. Place all foods into the trash. The iodine has to be reduced with a twofold excess of 50% sodium thiosulfate solution. Use 3 M sulfuric acid (H_2SO_4) to make the pH in the range 2 to 3. Set it aside for an hour. Neutralize the solution with sodium hydroxide (NaOH) or sulfuric acid and flush it down the drain with plenty of water. Evaporate the copper solutions to a solid that can be placed in a container to be buried in a secure landfill.

EXTRA
Sodium thiosulfate is also known as photographer's "hypo."

CARBOHYDRATES
PRE-LAB QUESTIONS

Name: _____

Lab Partner: _____

Section: _____   Date: _____

1) Which should provide energy more quickly to the body: table sugar or an apple that contains fructose? Explain your reasoning.

2) Which should provide energy more quickly to the body: sucrose or starch? Explain your reasoning.

3) What must be the active ingredient in the product Lactaid®, an aid for the digestion of dairy products containing lactose?

4) Give the results you would expect to see from each of the three chemical analysis tests—Benedict's, Barfoed's, and iodine—on the disaccharide maltose.

5) Which of the four carbohydrate solutions in step 7 would you predict to have a negative result to Benedict's test?

CARBOHYDRATES
REPORT SHEET

Name: _____

Lab Partner: _____

Section: _____ Date: _____

I. TESTS

Sample	Benedict's (Color)	Barfoed's (Color)	(Time)	Iodine (Color)	Conclusion (Type of Carbohydrate)
Glucose	_____	_____	_____	_____	_____
Fructose	_____	_____	_____	_____	_____
Sucrose	_____	_____	_____	_____	_____
Starch	_____	_____	_____	_____	_____
Cellulose	_____	_____	_____	_____	_____
Table sugar	_____	_____	_____	_____	_____
Corn syrup	_____	_____	_____	_____	_____
Honey	_____	_____	_____	_____	_____
Powdered sugar	_____	_____	_____	_____	_____
Sucrose in saliva	_____	_____	_____	_____	_____
Sugar in saliva	_____	_____	_____	_____	_____
Starch in saliva	_____	_____	_____	_____	_____
Cellulose in saliva	_____	_____	_____	_____	_____

II. ENZYME ACTION

Sample:	Starch	Sucrose	Table Sugar	Cellulose
Color:	_____	_____	_____	_____

III. QUESTIONS

1. From your data, which of the first 8 substances would you expect to provide the quickest energy to the body? Why?

2. Why does the water from the cooking of vegetables taste sweet?

3. In the procedure, heat was used to break the bond between monosaccharide units. In what other ways could the bonds be broken?

4. Give an explanation for the results in the enzyme action test.

INVESTIGATION

52 *Analysis of Fertilizer*
How Well Does Your Garden Grow?

OBJECTIVES
To use control solutions in each of several tests to determine the composition of fertilizer.
To perform a quantitative test for the phosphate ions and calculate the percent phosphate
in a fertilizer sample.

Relates to Chapter 20 of Chemistry for Changing Times, *twelfth ed.*

BACKGROUND
Agriculturists have found that not only more plants per acre but also larger plants can be
grown if the correct nutrients are added. The nutrients in soil are depleted by plants and need to
be replenished. The main nutrients plants need to obtain from the soil by are potassium, nitrogen,
and phosphorus. These nutrients are needed in usable forms such as potassium ions (K^+), nitrate
(NO_3^-) or ammonium (NH_4^+) ions, and phosphate ions (PO_4^{3-}). These nutrients can be replaced
by applying fertilizer to the soil. The amount of each nutrient is listed on a package of fertilizer
as a series of three numbers, such as 20-10-5. That means the fertilizer contains 20% N, enough
phosphorus to equal 10% P_2O_5, and enough potassium to equal 5% K_2O.

Most fertilizers contain the three primary nutrients and may also contain the secondary
nutrients—sulfur, calcium, and magnesium—and even some of the trace nutrients. Tests for
these elements have been developed and can be used to determine whether a fertilizer contains
the necessary ingredients.

Phosphate can be detected and also measured quantitatively by reacting phosphate ions with
the magnesium ion and ammonia to produce the precipitate $MgNH_4PO_4 \cdot 6H_2O$. This precipitate
can be filtered, dried, and weighed. The percentage of phosphate as P_2O_5 can then be calculated.

The reaction between iron(II) ions, nitrate ions, and sulfuric acid produces a complex ion
called nitrosyliron(II), $[Fe(NO)^{2+}]$, which causes a brown ring to form at the interface between
the sulfuric acid layer and the iron(II) sulfate layers. The nitrate ions remain in the sulfate layer.
Thus only the nitrate ions at the interface react with the sulfuric acid. The nitrogen in the nitrate
ion is reduced by the sulfuric acid from a +5 oxidation state to a +2 oxidation state.

Plants need nitrogen and are best able to use it in the nitrate form (NO_3^-). Ammonium and
ammonia in fertilizers are converted by bacteria in the soil to the nitrate ion. In the process, the
pH of the surrounding soil is lowered (made more acidic). The ammonium ion can be detected
by the production of ammonia gas, which is given off in the reaction:
$$NH_4^+ + OH^- \rightarrow NH_3(g) + H_2O.$$
The ammonia is detected by its distinctive odor or by the change in color of red litmus to blue.

The potassium ion can be detected by the presence of a violet flame in a flame test. The
wavelength of energy emitted as the electrons return to the original ground state energy level or
shell is that of the color violet. Sodium ions produce a bright yellow flame that can easily
overwhelm the delicate violet odor. A cobalt glass will filter out the yellow color.

Sulfate ions will precipitate with barium ions:
$$Ba^{2+} + SO_4^{2-} \rightarrow BaSO_4(s)$$

425

In this investigation, the tests are usually used first on a known solution and then on the fertilizer.

WASTE AND THE ENVIRONMENT

Barium is very toxic. Acidic or basic solutions can damage the plumbing if not neutralized or diluted.

⚠ *CAUTION*s warn about safety hazards.
*EXTRA*s give helpful hints, additional information, or interesting facts.

Reagents

sodium sulfate [Na_2SO_4]	10% sodium hydroxide [NaOH]
potassium chloride [KCl]	3 M hydrochloric acid [HCl]
ammonium phosphate [$(NH_4)_3PO_4$]	6 M hydrochloric acid [HCl]
sodium nitrate [$NaNO_3$]	12 M hydrochloric acid [HCl]
2 M ammonia [NH_3]	6 M nitric acid [HNO_3]
iron(II) sulfate solution 0.2M [$FeSO_4$]	75% isopropyl alcohol
concentrated sulfuric acid [H_2SO_4]	red litmus
10% magnesium sulfate heptahydrate [$MgSO_4 \cdot 7H_2O$]	
5% barium chloride [$BaCl_2$] or barium nitrate [$Ba(NO_3)_2$]	

Common Materials

inorganic fertilizer sample(s)

Special Equipment

platinum or nichrome wire	2 50 mL beakers
cobalt glass	250 mL beaker
filter paper	100 mL graduated cylinder
Buchner funnel	10 mL graduated cylinder
evaporating dish	balance
burner or hotplate	

PROCEDURE
Phosphate Test

1. Weigh between 3.0 and 3.5 g of the fertilizer to the nearest 0.01 g. Carefully transfer ALL the fertilizer to a 250 mL beaker. Add 40 mL of distilled water. Stir to dissolve the sample. If solids remain, filter the solution in a glass funnel lined with filter paper prepared as shown in the figure below.

EXTRA
Liquid plant food is simpler to use than a solid.

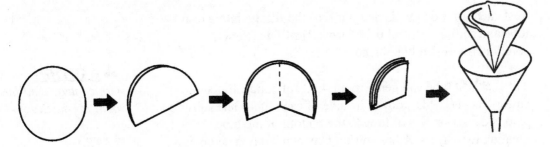

2. Add 45 mL of 10% magnesium sulfate heptahydrate ($MgSO_4 \cdot 7H_2O$) to the filtrate. While stirring gently, slowly add 150 mL of 2 M ammonia (NH_3). A white precipitate will form. Set the beaker aside for 15 minutes.

3. Filter the precipitate on a pre-weighed piece of filter paper in a Buchner funnel as shown in the figure below.

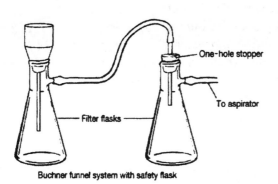

Buchner funnel system with safety flask

4. Rinse the beaker with two or three 5-mL portions of distilled water onto the precipitate.

5. Rinse the beaker with two 10 mL portions of 75% isopropyl alcohol onto the precipitate. Draw air through for 15 minutes to dry the precipitate. When dry, weigh and record the mass.

Nitrate Test

6. Dissolve 0.1 g of sodium nitrate ($NaNO_3$) in 15 mL of water to use as a known solution for the nitrate test. Dissolve 0.5 g of fertilizer in 15 mL of water as the fertilizer solution.

7. Filter the fertilizer solution through filter paper to remove solids as in step 1. The filtrate is the solution you will use.

8. Test each solution by placing 4 mL of the filtrate into a test tube and adding 2 mL of 0.2 M iron(II) sulfate ($FeSO_4$) solution. Swirl gently to blend.

9. Tilt each test tube and gently pour 4 mL of concentrated sulfuric acid (H_2SO_4) down the side of each tube. The sulfuric acid is denser, so it will flow to the bottom of the tube, creating two layers. A brown ring between the two layers is a positive test for the nitrate ion.

⚠ *CAUTION*
Concentrated sulfuric acid is caustic.

EXTRA
Don't shake the test tube.

Ammonium Ion

10. Use ammonium phosphate, $(NH_4)_3PO_4$, as the known compound. Test the known and the fertilizer by placing 2 g of each compound on separate evaporating dishes.

11. Gently warm and carefully sniff for the characteristic smell of ammonia.

⚠ *CAUTION*
Don't burn yourself.
⚠ *CAUTION*
Sodium hydroxide is caustic.

12. Add 5 mL of 10% sodium hydroxide (NaOH) to the evaporating dish. Place a moist piece of red litmus paper on the under side of a watch glass and set the watch glass on top of the evaporating dish. A blue color on the litmus paper is a positive test for the ammonium ion. If the litmus does not turn blue, warm the evaporating dish gently.

EXTRA
A false-positive test will occur if the NaOH spatters onto the litmus paper.

Potassium Test

13. Dissolve 0.1 g of potassium chloride (KCl) in 15 mL of 6 M hydrochloric acid (HCl) for the known solution. Dissolve 0.5 g of fertilizer in 15 mL of 6 M hydrochloric acid.

⚠ *CAUTION*
Hydrochloric acid is caustic.

14. Clean a platinum or nichrome wire by dipping it in concentrated (12 M) hydrochloric acid and burning off the moisture in a flame. Hold the tip of the wire just above the blue cone inside the flame just below the tip of the outer portion of the flame. Dip the wire in the acid and place it in the flame repeatedly until the wire burns clean, that is, the wire produces no added color to the flame.

⚠ *CAUTION*
Don't burn yourself.

15. To test the known solution, dip the wire into the solution and then place it in the flame. The yellow flame is due to sodium. To filter out the yellow component of the flame, look at the flame through a cobalt glass. The dark purple color of the glass will filter out the yellow sodium color but will allow you to see the violet color of potassium. A red-orange color may be due to calcium.

16. Clean the loop by repeating step 8 and then test the fertilizer solution as you did the known solution. If you do not see a positive test for potassium, heat the fertilizer solution to a boil for 20 seconds and try the flame test again.

Sulfate Test

17. Make a known solution for the sulfate test by dissolving 0.1 g of sodium sulfate (Na_2SO_4) in 10 mL of water.

18. To make the fertilizer solution, dissolve about 1 g of fertilizer in 15 mL of 6-M nitric acid (HNO_3). Boil gently for 10 minutes. Replace the lost volume with more acid.

19. Filter the solutions to remove the solids as in step 1.

20. Test each solution by adding 2 mL of 5% barium chloride ($BaCl_2$) or barium nitrate, $Ba(NO_3)_2$, solution.

21. Filter each solution through filter paper to separate the solids from the solutions.

22. Wash the precipitates with 5 mL of water several times. If the precipitates do not dissolve, wash them with 5-mL portions of 3-M hydrochloric acid. If a precipitate is not soluble in water or in hydrochloric acid, it is barium sulfate and is an indication of the presence of sulfate ion.

23. Place the barium sulfate in a waste receptacle to go to the toxic waste site. Flush the solutions down the drain with plenty of water. Throw the phosphate precipitate into the trash.

⚠ _CAUTION_
Hydrochloric acid is caustic.

ANALYSIS OF FERTILIZER
PRE-LAB QUESTIONS

Name: _____

Lab Partner: _____

Section: _____   Date: _____

1) What would be the content of a fertilizer that is labeled 20:20:10?

2) Some plants such as conifers and other evergreens need a low soil pH. Which number would you want to be the higher number on the fertilizer package for these plants?

3) Flowering plants use different nutrients from the soil than does a lawn. If a lawn food is 15-30-15, but a food for flowering plants is 30-10-10, which nutrient is most needed for each type of plant?

4) Some fertilizers are labeled "all-purpose." What would you expect the relative nutrient content to be in such a fertilizer?

ANALYSIS OF FERTILIZER
REPORT SHEET

Name: _____

Lab Partner: _____

Section: _____   Date: _____

I . Describe the result of each test on the known solutions

    Nitrate

    Ammonium

    Potassium

    Sulfate

II . FERTILIZER TEST

	Present	Absent
Phosphate	_____	_____
Nitrate	_____	_____
Ammonium	_____	_____
Potassium	_____	_____
Sulfate	_____	_____

III. Mass of fertilizer phosphate precipitate and paper   _____ g

    Mass of paper   −_____ g

    Mass of fertilizer phosphate precipitate ($MgNH_4PO_4 \cdot 6H_2O$)   _____ g

$$\underline{\hspace{2cm}} \text{ g MgNH}_4\text{PO}_4 \bullet 6\text{H}_2\text{O} \quad \times \quad \frac{1 \text{ mol MgNH}_4\text{PO}_4 \bullet 6\text{H}_2\text{O}}{245.3 \text{ g MgNH}_4\text{PO}_4 \bullet 6\text{H}_2\text{O}}$$

$$\times \quad \frac{1 \text{ mol P}_2\text{O}_5}{2 \text{ mol MgNH}_4\text{PO}_4 \bullet 6\text{H}_2\text{O}} \times \frac{141.9 \text{ g P}_2\text{O}_5}{1 \text{ mol P}_2\text{O}_5} = \quad \underline{\hspace{2cm}} \text{ g P}_2\text{O}_5$$

$$\underline{\hspace{2cm}} \frac{\text{g P}_2\text{O}_5}{\text{g fertilizer}} \times 100 \% \quad = \quad \underline{\hspace{1.5cm}} \% \text{ P}_2\text{O}_5$$

IV. QUESTIONS

1. Arsenic reacts chemically like phosphorus. Why should fertilizer not contain arsenic?

2. Liquid fertilizers often contain the same ions as in solid fertilizer, only they are dissolved in water. Will the tests in this investigation work on liquid fertilizers?

3. Which fertilizer would be needed if the soil were poor in nitrogen: a 10-10-10 or a 20-5-5?

4. How did your measured percent phosphate compare with the amount indicated on the label of the fertilizer container?

5. A popular fertilizer for vegetable gardens is listed as 18-18-21. What does this ratio say about the relative importance of each nutrient to vegetable plants?

53 *Saponification*

Lard to Lye Soap

OBJECTIVES

To understand the process of saponification.
To produce a small amount of soap from cottonseed oil and sodium hydroxide.
To calculate the average molar mass of soap produced from cottonseed oil.

Relates to Chapters 21 and 9 of Chemistry for Changing Times, *twelfth ed.*

BACKGROUND

In the times of our grandparents and great-grandparents, a wide range of cleaning products was not available. The main cleaning product was lye soap. It was used for a large number of cleaning jobs: bathing, hair washing, clothes washing, and dishwashing. It was usually formed into a bar, but it could be cut into chips to make it more soluble for washing dishes or clothes.

Lye soap was made by reacting a fat or oil with lye or sodium hydroxide [NaOH], also called caustic soda. Although many soaps are now made with coconut oil or palm oil, animal fat was more commonly used in the days of our grandparents. Some restaurants even use the grease collected in the kitchen to make soap to wash the dishes. Potassium hydroxide can be used instead of sodium hydroxide.

The soap-making process is called *saponification*. In the reaction, a fat and sodium hydroxide react to produce soap with glycerol as a by-product. Before the reaction, the oils and fats are in a form we call *triesters* (three ester groups). The reaction breaks the ester group apart so that three chains ending in a carboxylate ion (COO⁻) and an alcohol group (HO) are formed. The compound containing the alcohol groups is *glycerol*. Glycerol is commonly separated and used in more expensive cosmetic applications, or it can be left in the soap and sold for a higher price. An example is expensive glycerin-containing shower gels. Evaporating part of the water from the solution allows the carboxylate ions to combine with the sodium ions to form a solid that is called *soap*.

After the reaction is complete, the soap is separated from the glycerol by "salting out." The solution is mixed with a concentrated sodium chloride solution. This electrolyte causes the dispersed soap to coagulate. The soap is then washed several times with concentrated sodium chloride to remove the excess lye.

The "R" in the diagram represents a straight-chain hydrocarbon of 12–20 carbon atoms. The three R groups are often different from one another. In cottonseed oil, there are three main R groups: linoleate, oleate, and palmitate. Each R group produces one of the three most abundant soap salts derived from cottonseed oil: sodium linoleate (approx. 50%), sodium oleate (approx. 30%), and sodium palmitate (approx. 20%).

$$CH_3CH_2CH{=}CHCH_2CH{=}CH(CH_2)_7{-}\overset{\displaystyle O}{\overset{\displaystyle \|}{C}}{-}ONa$$

Sodium Linoleate

$$CH_3(CH_2)_7CH{=}CH(CH_2)_7{-}\overset{\displaystyle O}{\overset{\displaystyle \|}{C}}{-}ONa$$

Sodium Oleate

$$CH_3(CH_2)_{14}{-}\overset{\displaystyle O}{\overset{\displaystyle \|}{C}}{-}ONa$$

Sodium Palmitate

Soap-making has recently made a comeback in the last few years. People desire "natural" materials and lye soap qualifies. Lye soap can be made from several different animal fats and vegetable oils. Each fat or oil has slightly different qualities. The soaps can be colored or scented or made soft through different treatments or additives. Additives can also change other properties of the soap. Ground-up pumice can make a harsh abrasive soap. An abrasive soap that is not so harsh can be made by adding uncooked oatmeal to the soap.

WASTE AND THE ENVIRONMENT

The compounds in this investigation are not toxic. Acidic and basic solutions can harm the plumbing if not neutralized or diluted.

▲ *CAUTION*s warn about safety hazards.

*EXTRA*s give helpful hints, additional information, or interesting facts.

Reagents
  20% sodium hydroxide solution [NaOH]
  sodium chloride [NaCl]
  calcium or magnesium salt solution
  pH paper

Common Materials
  cottonseed oil
  scented oil

Laboratory Equipment
  150 mL beaker
  100 mL graduated cylinder
  10 mL graduated cylinder
  stirring rod
  filter paper
  ring stand, ring, and wire gauze
  laboratory burner
  lab gloves

PROCEDURE

1. Place 10 mL of cottonseed oil and 15 mL of 20% sodium hydroxide (NaOH) in a 100- or 150 mL beaker above a laboratory burner. If it is available, a hot plate is safer to use. Heat the mixture strongly with vigorous stirring. The stirring is necessary or the reactants will remain in two layers. The mixture will begin to foam and rise in the beaker. Continue to heat and stir. When the foam dies down, the mixture will appear syrupy. Continue to heat and stir. Vigorous stirring with a long stirring rod (at least 8 in.) should prevent spattering and frothing as long as a slow, controlled boiling is maintained. The long rod is to keep your hand far from the beaker. Keep your head back and below the opening of the beaker. When the mixture is waxy, discontinue heating. Waxy means that it sticks to and builds up on the stirring rod. The mixture should still appear moist or it has been overcooked. The boiling has removed all of the water from the mixture. One purpose of the stirring is to prevent the soap from forming a solid on the bottom of the beaker. A solid layer on the bottom will often char.

2. When all the water has apparently been removed, let the mixture cool slightly. If a waxy solid forms, the process is complete. Let the mixture cool. If a syrupy liquid results, the reaction is not complete, and heating must be resumed.

3. While the soap is cooling, make up a concentrated sodium chloride (NaCl, also called table salt) solution by adding 18 g of NaCl to 60 mL of water.

4. Pour 20 mL of the concentrated sodium chloride solution into the beaker containing the soap. Break up the soap into small pieces with the stirring rod so that all the soap is washed free of glycerol and sodium hydroxide.

5. Filter the soap solution with a piece of filter paper in a funnel as shown on the next page. Fold a piece of filter paper in half and in half again. Tear off the outer corner of the fold. Open the paper between the first and second folds, and place the paper in the funnel. Moisten the filter with water so the filter paper clings to the funnel. Pour the soap mixture through the filter paper, collecting the filtrate in a beaker.

⚠ *CAUTION*
The stirring rod can cause the beaker to overturn.

Don't burn yourself.

Sodium hydroxide is caustic.

⚠ *CAUTION*
Don't handle the soap, as it still contains caustic NaOH.

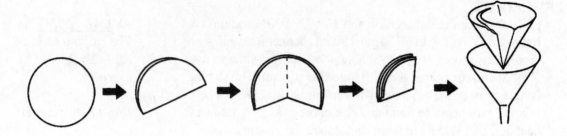

6. Scrape the soap back into the beaker and repeat the washing with each of the other two 20 mL portions of concentrated sodium chloride.

7. While wearing gloves, work the soap on a piece of dry filter paper or paper toweling to remove the last part of the wash water. Weigh the soap.

8. **Optional Step**
   To add scent to the soap, place it in a beaker. Add a few drops of scented oil. Mix thoroughly with a stirring rod.

9. Press the soap into a block. Show the soap to your instructor for approval.

EXTRA
If the soap was to be molded, this would be the time to do it.

10. Add a piece of soap about the size of a pea to 5 mL of water in a test tube. Stopper tightly and shake vigorously. Use pH paper to determine the pH of the solution.

11. Wash your hands with a small piece of your soap.

⚠ *CAUTION*
If the soap is very alkaline (pH>11), wash quickly. Rinse with plenty of water.

12. Place 5 mL of distilled water in a test tube. Place 4 mL of tap water and 1 mL of a calcium or magnesium salt solution into another test tube. Put a small piece of soap into each test tube and shake vigorously. Now compare the lathering in both tubes. Look for a gelatinous precipitate in each test tube.

EXTRA
Calcium or magnesium ions cause water to be hard.

13. Throw the solids in the trash. Flush the solutions down the drain with plenty of water.

SAPONIFICATION
PRE-LAB QUESTIONS

Name: _____

Lab Partner: _____

Section: _____   Date: _____

1) Baths have not always been comfortable. Using the information about "grandma's" soap, give several reasons why that was so.

2) Why do the two reactants in step 1 remain in separate layers unless stirred?

3) Why should scent be added to the soap after it is complete but before it is pressed into a cake or molded?

4) Soap that is ground into flakes or tiny pellets is used as dishwasher detergent or laundry soap. Would these types of soap be cooked a longer or shorter time than the soap you have made? Why do you think so?

SAPONIFICATION
REPORT SHEET

Name: _____

Lab Partner: _____

Section: _____   Date: _____

I. MASS OF SOAP _____

    Instructor's approval of soap   _____

II. QUESTIONS
  1. Explain why you washed the soap with salt solution rather than with water.

  2. What was the pH of soap in water?

  3. How do you explain this pH?

  4. Describe any observations about washing your hands with a small piece of the soap.

  5. In step 12, which aqueous solution allowed the soap to lather more freely?

6. The three most abundant soap molecules produced from cottonseed oil are sodium linoleate, often written as $C_{18}H_{31}O_2Na$; sodium oleate, written as $C_{18}H_{33}O_2Na$; and sodium palmitate, written as $C_{16}H_{31}O_2Na$. The average soap molecule is about 300 g/mol. The cottonseed oil would consist of the three soap molecules as fatty acids combined on a three-carbon backbone. Cottonseed oil would be about 872 g/mol. Calculate the expected yield of soap using the cottonseed oil as the limiting reagent. The density of cottonseed oil is 0.92 g/mL.

$$10 \text{ mL oil} \times \frac{0.92 \text{ g}}{\text{mL}} \times \frac{\text{mole cottonseed oil}}{872 \text{ g cottonseed oil}} \times \frac{3 \text{ mol soap}}{\text{mol soap}} \times \frac{300 \text{ g soap}}{\text{mole soap}} = \underline{\hspace{1cm}} \text{ g soap}$$

7. How does actual weight compare with calculated yield? Does the soap still contain water?

54 *Soap in Hard Water*

To Lather or Not to Lather

OBJECTIVES

To comprehend the role "hard-water" ions play in the effectiveness of soap and detergent.

To gather quantitative data on the lathering ability of soap and detergent in water of varying "hard-water" ion concentrations.

To understand the source of soap scum and the role of a builder or water softener in preventing its formation and increasing the effectiveness of cleaning agents.

Relates to Chapter 21 of Chemistry for Changing Times, *twelfth ed.*

BACKGROUND

Water may contain a large number of different ions. When water contains ions of Ca^{2+}, Mg^{2+}, and/or Fe^{2+} in the concentration range of 125 to 250 ppm (125 to 250 mg ions per liter), it is called "hard" water. These ions react with ion in the soap to form an insoluble precipitate as illustrated in the following reaction:

$$2\ C_{17}H_{35}COO^-Na^+ \quad + \quad Ca^{2+} \longrightarrow \quad Ca(C_{17}H_{35}COO)_2 \quad + \quad 2\ Na^+.$$

sodium stearate soap	ions in "hard water"	solid "soap scum"

This precipitate is often a sticky, slimy substance that may cling to clothes, causing dinginess or "telltale gray." Part of the precipitate floats on the water, forming a scum. After a bath, the substance clings to the surface of the tub and is known as "bathtub ring." Washing in hard water costs more money because part of the soap is tied up as a precipitate with Ca^{2+}, Mg^{2+}, and Fe^{2+} ions. Extra soap is also required to wash away the precipitate. It is an unusual fact that some condiments contain calcium stearate and it is used as an additive to prevent caking in garlic salt, vanillin powder, salad-dressing mixes, meat tenderizers, mints, and some candies. The stearate ion commonly appears in soaps and in the bathtub, calcium stearate would be soap scum.

Soaps are called *surfactants* because they are surface-active agents. Soaps will work at the surface of an object to separate the dirt from the material. The nonpolar carbon-chain end of the soap molecule dissolves in the oily component of the dirt. The ionic end of the soap then sticks out of the dirt forming a ball, or *micelle*, as the mass of dirt is surrounded by imbedded soap molecules. Water molecules are attracted to the ionic end of the soap molecule, surround the micelle, and carry away the dirt.

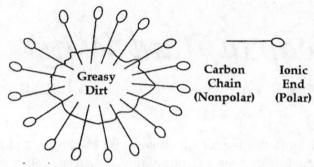

Builders are chemicals added to soaps and detergents to tie up, or precipitate, the hard-water ions. They also control the alkalinity and in some way increase the ability of surfactants to keep dirt suspended in water. Some of these builders are sodium carbonate, borax, sodium sulfate, sodium silicate, and sodium metaphosphate. The phosphates were very popular until the effects on the environment became evident. Phosphates increase the growth of algae, causing the dissolved oxygen in the water to be depleted. Waterways in which high concentrations of phosphates are dumped can become foul-smelling, anaerobic, swamp-like bodies of water. Fish and other types of clean aquatic life cannot survive in the *anaerobic*, or oxygenless, environment. Many states have now imposed various restrictions on the use of phosphates.

Detergents are synthetic soaps made from petroleum and like soap they still contain a hydrophobic (water-fearing) end and a hydrophilic (water-loving) end. Because of the two different ends on the detergent molecule, it acts like a soap in that it will hold together molecules of oil and water. In this way, the oil is held suspended in the water. Most surfactants will stabilize suspensions of oil and water. One of the more common types of detergents is the LAS type of molecule:

$$H_3C(H_2C)n-\!\!\!\!\bigcirc\!\!\!\!-SO_3^-$$

hydrophobic end   hydrophilic end
Linear **A**lkyl Benzene **S**ulfonate

WASTE AND THE ENVIRONMENT
The compounds used in this investigation are not toxic.

⚠ *CAUTION*s warn about safety hazards.
*EXTRA*s give helpful hints, additional information, or interesting facts.

Reagents
    calcium nitrate [$Ca(NO_3)_2$]   sodium meta-phosphate [$NaPO_3$]
    sodium carbonate [Na_2CO_3]   distilled water
Common Materials
    Ivory Snow® or castille soap   red food coloring
    Tide®   borax
Laboratory Equipment
    filter paper   5 125 mL Erlenmeyer flask
    250 mL beaker   laboratory burner
    2 1000 mL beaker   ring stand, ring, and wire gauze
    eyedropper   stirring rod

PROCEDURE

1. **Preparation of Soap Solution**
   One pair of students will prepare the soap solution for two pairs of students. Place 500 mL of distilled water in a 1000 mL beaker. Add 2.5 g of Ivory Snow® to the beaker. Place the beaker on a ring stand and heat gently until all of the soap dissolves. Set the beaker aside to cool while you prepare the "hard-water" solution. When the 1000 mL beaker has cooled enough to handle, filter the soap solution.

⚠ CAUTION
Don't burn yourself.

2. Fold a piece of filter paper in half and in half again. Tear off the corner of the outer fold. Open up the paper between the first and second folds and place the paper in the funnel. Moisten the paper with distilled water so that the filter paper adheres to the funnel. Slowly filter the soap solution through the filter paper into a 1000 mL beaker.

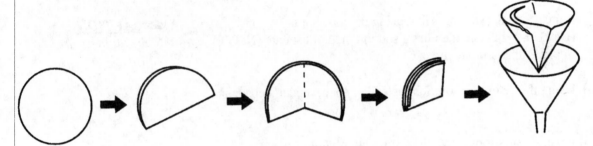

3. Pour half the solution into each of two beakers. Mark on the beakers "soap solution" and set one aside. Give the other beaker to the other pair of students.

4. **Preparation of Detergent Solution**
   The other pair of students will prepare the detergent solution for both pairs of students. Prepare the detergent solution just like the soap solution except use 1.0 g of Tide® in 200 mL of distilled water.

5. **Preparation of Hard Water**
   Prepare "hard water" by adding 0.56 g of calcium nitrate [Ca(NO$_3$)$_2$] to 1 L of distilled water. Stir with a clean stirring rod until all solid is dissolved. Place 125 mL of the hard water in a 250 mL beaker. Label this beaker as "hard water" and set it aside.

6. **Testing for Lather**

   Place about 25 mL of distilled water in a 125 mL Erlenmeyer flask and add 1 drop of red food coloring. Add the soap solution by dropping a few drops at a time from an eyedropper into the distilled water in the flask. Agitate the flask vigorously by swirling it after each addition. Continue to drop soap and agitate until a lather that persists for at least 30 seconds covers the entire surface of the solution in the flask. Count the number of drops added and record them in the report sheet.

   EXTRA
   Only soap added after the lather persists is available for cleaning.

7. Repeat step 6 using first tap water and then the hard water that you prepared.

8. Pour 25 mL of hard water into a 125-mL Erlenmeyer flask. Add 0.1 g of sodium carbonate (Na_2CO_3) and stir. Allow the solution to sit for 10 minutes. Repeat step 6 using the chemically treated "hard water" instead of distilled water.

   EXTRA
   To save time, set up step 9 during the 10 minutes.

9. Repeat step 8 two times, first using 0.4 g of borax ($Na_2B_4O_7) \cdot H_2O$ and then 0.2 g sodium metaphosphate ($NaPO_3$) in place of sodium carbonate.

   ⚠ *CAUTION*
   Some builders are caustic.

10. Using the detergent solution instead of soap, repeat steps 6 and 7.

11. Flush all solutions down the drain with plenty of water.

SOAP IN HARD WATER
PRE-LAB QUESTIONS

Name: _____

Lab Partner: _____

Section: _____   Date: _____

1) List another substance that might qualify to be classified as a surfactant besides soap and detergent.

2) Have you ever had the experience of washing your hands or hair and had the sensation that you could not get the soap off? What would you suspect about the metallic ion content of the water?

3) Dry cleaning is a process in which no water is used. The cleaners in the process still attract the dirt, but do not need to be attracted to polar water molecules. Describe the probable structure of a molecule of the dry cleaning agent.

4) People who rely on septic tanks for their sewage disposal are ultimately putting all of the waste water from their homes into the environment of the immediate area. Which substances would be harmful to use? Is it your assessment that they should rely on detergent or soaps more heavily? Why?

SOAP IN HARD WATER
REPORT SHEET

Name: _____

Lab Partner: _____

Section: _____   Date: _____

I. SOAP AND DISTILLED WATER
of drops to lather   _____

Volume = # of drops × 0.05 mL/drop = _____ mL

II. SOAP AND TAP WATER
of drops to lather   _____

Volume = # of drops × 0.05 mL/drop = _____ mL

III. SOAP AND HARD WATER
of drops to lather   _____

Volume = # of drops × 0.05 mL/drop = _____ mL

IV. SOAP AND HARD WATER WITH Na_2CO_3

of drops to lather   _____

Volume = # of drops × 0.05 mL/drop = _____ mL

V. SOAP AND HARD WATER WITH BORAX

of drops to lather   _____

Volume = # of drops × 0.05 mL/drop = _____ mL

VI. SOAP AND HARD WATER WITH $NaPO_3$

of drops to lather   _____

Volume = # of drops × 0.05 mL/drop = _____ mL

VII. DETERGENT AND DISTILLED WATER

of drops to lather   _____

Volume = # of drops × 0.05 mL/drop = _____ mL

VIII. DETERGENT AND TAP WATER

of drops to lather   _____

Volume = # of drops × 0.05 mL/drop = _____ mL

IX. DETERGENT AND HARD WATER

of drops to lather   _____

Volume = # of drops × 0.05 mL/drop = _____ mL

X. QUESTIONS

1. Why is there a difference between distilled water and tap water in the volume of soap required to achieve a sustained lather? Between tap water and hard water?

2. Explain why the addition of a builder alters the amount of soap required to produce a sustained lather.

3. Which is the most effective builder encountered in this investigation?

4. Why do clothes washed in hard water get cleaner when a builder is added?

5. How is detergent different from soap in its ability to sustain a lather?

INVESTIGATION

55 *Personal Products*

Smell Good and Look Nice!

OBJECTIVES

To separate the different notes or scents that make up a cologne, perfume, or aftershave.

To observe the solvent action of different cleaning agents on lipstick.

To demonstrate the polymeric properties of hair spray.

Relates to Chapter 21 for Chemistry for Changing *Times, twelfth ed.*

BACKGROUND

The field of personal care is another area in which many people practice chemistry. Not only personal-care producers, but also product users—people like you—become practicing chemists.

Cosmetics, which belong to one category of personal-care products, have become a very large industry in the United States and are used by many people in order to be more physically appealing. Some people would never let others see them without their "face" on.

Many facial cosmetics are made of a base of semisolid wax and oil that also contains a coloring agent. Advertising makes each brand appear to have marvelous, wonderful properties, but the important ingredients are the coloring agents carried by one or more waxes and/or oils.

Lipsticks are used to impart color to the lips. Most lipsticks impart the same color that appears in the tube, but some specialty lipsticks show up as a different color, depending upon the chemicals encountered on the body and/or the temperature of the body.

Removing makeup is another chemical process that is practiced by many people. The key to removing makeup is to dissolve the material. "Like dissolves like" is a good simple rule to remember when removing makeup. Polar compounds are best removed by polar solvents, while nonpolar substances, such as waxes and oils, are removed most efficiently by nonpolar solvents. Some of the materials you will test as agents for removing makeup are water, soap and water, cleansing or cold cream, baby oil, and an astringent.

Water is an excellent solvent for polar and ionic compounds and small compounds containing nitrogen, oxygen, and fluorine. Soap is soluble in water and in long-chain organic molecules. Soap will surround a small ball of grease or oil, allowing it to be carried away by water. Cleansing cream is composed of long-chain organic compounds. It is designed to dissolve oils and fats and then be wiped away by a cloth or tissue. Baby oil is composed of smaller organic molecules that are liquid. Often less expensive than cleansing cream, baby oil is sometimes used for removal of very dry or heavy makeup, such as the type worn by actors. An astringent is not a cleanser; it is to be used after cleansing, and causes the pores to close so that dirt will not enter the pore.

Many cosmetics contain materials that smell good, often to hide offensive odors. Pleasant-smelling compounds, both natural and synthetic, are combined to make an aroma that is pleasing initially and continues to be pleasing. The perfumes that people wear are a combination of scents in an alcohol solvent. Colognes are about one-tenth as concentrated as perfumes. The scents that comprise a perfume, cologne, or aftershave are divided into three groups called *notes*. The top note is made of lighter or more *volatile* materials that evaporate within a few minutes of application. The middle note, which may last around thirty minutes, is composed of less volatile compounds. They take longer to evaporate and therefore last longer. The bottom note is the slightly volatile, long-lasting component of perfumes. It is often derived from animal *musk*, which is a richly scented substance secreted by animals and is present in most soaps, shampoos, powders, cosmetics, bath oils, household cleansers, laundry detergents, and even in some foods—in fact, it can be found in nearly everything that requires a fragrance.

The boiling stones used in this investigation allow gas bubbles formed during boiling to remain small and to form on a continual basis rather than in sudden bursts. Without the stones, the liquid could become superheated. Also, when a gas bubble does form, it is often large and can cause the liquid to spatter out of the container.

To keep a hairdo in place, many people use hair spray. Most hair sprays are solutions of polymers dissolved in a volatile solvent. When hair spray is sprayed onto hair, the solvent evaporates, leaving a resin to hold the hair in place. A few hair sprays use the same polymer that is the base for some chewing gums.

In this investigation, you are going to investigate makeup removal, the polymeric effects of hair spray, and separation of the notes of a perfume.

WASTE AND THE ENVIRONMENT
The compounds used in this experiment are not toxic.

⚠ *CAUTION*s warn about safety hazards.
*EXTRA*s give helpful hints, additional information, or interesting facts.

Common Materials
    lipstick
    cleansing cream
    bar hand soap
    baby oil
    astringent
    cotton balls
    cologne or perfume
    hair spray
    paper towels

Laboratory Equipment
    25 mL vials with stoppers or caps
    boiling stones (marble chips)
    watch glass
    small test tube
    3 vials with caps or stoppers

PROCEDURE

1. **Lipstick Removal**
   On the inside of your forearm, mark five stripes with a lipstick. Compare the color of your stripes with that of others who used the same lipstick.

2. Attempt to remove each of the stripes one at a time with each of following five substances using a paper towel: water, soap and water, cleansing cream, baby oil, and astringent. Compare the areas of the five stripes and answer the report questions.

3. Pour 1 mL of liquid hair spray onto a watch glass. Set it aside to dry.

4. Place 6 mL of cologne in a small test tube and add one or two boiling stones. Place a cotton ball in the top of the test tube and place the test tube in a hot-water bath. The cotton ball should not touch the liquid surface. Keep the water bath below 60°C.

5. When the volume is reduced by one-third, remove the cotton ball with forceps and place the cotton ball in a vial marked "top note" and stopper the vial.

6. Place a new cotton ball in the top of the test tube and continue heating until the volume is one-third of the original volume. Remove the cotton ball and place the cotton ball in a vial marked "middle" and stopper the vial.

7. Place a new cotton ball in the top of the test tube and continue heating the test tube until the volume is less than 1 mL. Place the cotton ball into a vial marked "bottom note" and stopper the vial.

8. After the vials have cooled for a few minutes, take them to an area where the perfume smell is not strong. One at a time, open and smell each vial. Close each vial before opening another.

9. Gently scratch loose one edge of the dried hair spray. Peel off the hair spray in one piece if possible. The center may be the easiest place to get loose because that is where it is the thickest. Test the film for stretch, flexibility, and strength.

EXTRA
Guys usually get lipstick secondhand. Now they'll learn how to remove it.

EXTRA
Afterward you may want to clean the five areas with the best cleaner.

⚠ *CAUTION*
Don't burn yourself. Don't point the test tube at anyone. Don't heat it too strongly. The alcohol solvent is flammable.

⚠ *CAUTION*
Don't burn yourself.

EXTRA
Each note is different.

EXTRA
The film is something like plastic wrap.

10. Flush all the solutions down the drain with plenty of water. Throw the cotton balls and dried hair spray into the trash.

PERSONAL PRODUCTS
PRE-LAB QUESTIONS

Name: _____

Lab Partner: _____

Section: _____   Date: _____

1) Heavy make-up called *grease paint* is used in theater applications, by the military, and by athletes. It does not wash off easily, many times not even with soap. Why not?

2) Why is it an advantage to the wearer that water does not remove grease paint?

3) Should an astringent be used before or after removing cosmetic substances from the skin? Why?

4) It is often said that hair spray is highly flammable. Would you expect this to be true when it is dry, or only when it is still wet? Explain.

PERSONAL PRODUCTS
REPORT SHEET

Name: _____

Lab Partner: _____

Section: _____   Date: _____

I. QUESTIONS

1. Are the stripes made by the lipstick on your arm the same color as the lipstick itself?

2. Are all of the stripes made by the same lipstick the same color on everyone's arm?

3. Do some stripes look moister?

4. Which cleanser dissolved the lipstick best?

5. Were any of the cleansers unable to clean the lipstick off your arm?

6. Did any cleanser leave a residue of itself?

7. Are the pores in the astringent area smaller?

8. Which cleanser would you use to remove lipstick?

9. Do the three notes smell alike?

10. Does one of the notes emit an unpleasant smell?

11. Describe each note:

Top

Middle

Bottom

12. Is the hair-spray film flexible?

13. Is the hair-spray film sticky?

14. Is the hair-spray film stretchable?

56 *Hair Chemistry*

The Long and Short on Color and Curls

OBJECTIVES

To understand the basic structure and composition of hair.
To learn the function of pH in different hair-care products.
To observe the changes that are made in hair under different pH conditions.

Relates to Chapter 21, 4, and 16 for Chemistry for Changing *Times, twelfth ed.*

BACKGROUND

Hair is a characteristic of mammals that serves several functions ranging from protective covering to adornment. Many modern-day humans spend a large amount of time and money on the appearance of their hair. This investigation will demonstrate the effects of changing pH on the structure and appearance of hair.

Protein is the substance that forms the structural part of living tissues. Proteins are composed mainly of the elements carbon, hydrogen, oxygen, nitrogen, with small amounts of phosphorus and sulfur. Hair is composed primarily of a protein called *keratin*. This protein comes in a hard form and a soft form. The soft form is familiar as calluses that form on hands and feet in areas exposed to large amounts of friction. The calluses function to reduce wear. The hard form of keratin is found in tissues such as fingernails and hair. This form of protein contains more sulfur than other forms. The sulfur allows for more cross-linking of the protein fibers and therefore adds strength. For this reason, hair is resistant to decay, decomposition, and digestion. Hair is found intact in archeological sites where all other protein tissues are long since disappeared.

A strand of hair is made of three layers, the cuticle, the cortex, and the medulla. Each layer has a specific composition. The outer layer, or *cuticle*, is formed of overlapping scales that are tough and waterproof. These scales protect the hair and prevent it from drying out. They also provide strength while at the same time their arrangement allows for flexibility. The scales hold the natural oils that give hair luster and keep the inner cuticle moist and healthy. Some hair types have very tight, close scales while other types of hair have loose, open scales. The loose, open arrangement allows for easier treatment of the hair.

As much as 90% of the hair strand lies in the middle layer, the *cortex*. This is the area where hair color resides. *Melanin* is the pigment found in human skin. The amount of melanin that is deposited in the cortex as it grows from the follicle determines the color of the hair strand. The cortex is formed of millions of protein fibers in a linear arrangement. In places where these fibers twist around each other, the hair has greater strength.

The inner section of a hair strand is called the *medulla*. The medulla is very small in relation to the other layers and its function is not well established.

Changing hair color and shape can be temporary or permanent. Temporary changes usually wash out because they are water-soluble. Any permanent change to hair color or curl requires a change beneath the cuticle layer and is potentially damaging to the hair.

Alkaline substances have a pH greater than 7 and thus are basic. Alkaline substances cause the cuticle to swell and open, allowing chemicals to enter the cortex. Acidic substances have pH levels lower than 7 and tend to harden the cuticle.

Hair colors can be temporary, semi-permanent, or permanent. Temporary hair colors are usually slightly acidic and reside only on the surface of the cuticle. These wash out. Semi-permanent colors are slightly alkaline and place color molecules into the cuticle layer. These eventually wash out because most soaps and shampoos are very slightly basic and cause the cuticle to swell a bit. Permanent hair color is applied in a chemical process that produces large colored molecules inside the cortex layer. These trapped molecules are chemically stable and are not water-soluble; therefore they are not removed by shampoo.

The natural color of hair can be removed with strong alkaline solutions like chlorine bleach; however, the substance of choice for professionals is usually a 10 or 15% hydrogen peroxide solution. The hydrogen peroxide at the grocery store is a 3% solution and can be used to lighten the shade of the hair, but will not bleach the color from the hair completely.

The shape of a hair strand is determined by a combination of hydrogen bonds and sulfide linkages between protein strands in the cortex. Hydrogen bonds are broken when the hair is wet. Therefore, hair can be "set" to dry and the new shape holds until the hair becomes wet again. Hydrogen bonds can also be altered with heat; thus, hot rollers and curling irons can put a temporary curl into a strand of hair.

Permanent curls and straightening of curly hair are the result of a chemical process in which the cuticle of the hair is "raised" using an alkaline treatment that also breaks the sulfide bonds in the cortex. The hair is treated while wrapped around a cylinder the desired size and then the hair is neutralized with an acid to return the pH to normal. The sulfide bonds that reform are covalent in nature and therefore will not dissolve in water. The permanent curl will "take" better if the hair is not shampooed for 48 hours following the chemical process.

Processes that alter the pH outside of a specific range can permanently damage hair. Any substances with a pH lower than 3 or greater than 10 have an adverse effect on the structure and condition of the hair. Many times hair that has undergone several permanent treatments is damaged to the point that it becomes brittle and breaks off easily. The hair can also become devoid of the ability to hold the natural oils that keep it healthy. To help counteract these effects, conditioners that contain processed proteins can be applied to fill in gaps where the protein of the cuticle is damaged or missing. This is a very temporary cure and must be reapplied after each shampoo.

Special note: The processes in this investigation are for demonstration purposes only and should not be used for personal hair care. They are potentially very damaging to the hair.

WASTE AND THE ENVIRONMENT

The solutions used in this investigation are not toxic, but many are caustic. The pH of the substances can be brought closer to neutral by diluting them with a large quantity of water before disposal.

⚠ *CAUTION*s will appear to warn about safety hazards.
*EXTRA*s will appear to indicate helpful hints, additional information, or interesting facts.

Common Materials
ammonia cleanser
vinegar
bleach
food color (blue is best)

shampoo
clear tape
2 dozen strands of long straight hair (12 light
    and 12 dark)

Laboratory Equipment

    100 mL graduated cylinder   stirring rod
    3 400 mL beakers   watch glass
    Extra stirring rods might speed up the investigation.

PROCEDURE
Gathering Strands of Hair
From yourself, classmates, or friends, gather 2 dozen strands of long, straight hair — 1 dozen strands of dark-colored hair, and 1 dozen strands of light-colored hair.

Permanent Curl
1. Place 90 mL of clear ammonia cleanser and 175 mL of water in a clean, dry 400 mL beaker. This produces a basic solution.

2. Wrap 6 strands of the long straight hair around a stirring rod or plastic pen cylinder in a spiral fashion. Use clear tape to secure each end of the hair to the ends of the rod. Submerge the coiled hair in the basic solution. Save the other 6 strands to use as a control for comparison purposes.

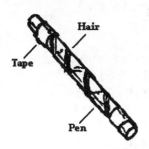

3. Remove the hair coil from the ammonia solution after it has soaked for 15 minutes. Leave the hair taped in place as you rinse it under tap water for 1 minute. Save the basic solution for use in a later step.

4. Place 225 mL of vinegar in a second 400 mL beaker. Submerge the rinsed hair from the ammonia solution in the vinegar. Allow it to sit for 5 minutes, then remove it and rinse it under tap water for 1 minute. Allow the hair to dry.

5. Carefully remove the tape from the hair and uncoil it. Make observations as you compare the coiled hair to the strands saved as a control.

EXTRA
Freshly shampooed hair with no spray, conditioners, or gels will work best.

EXTRA
The basic solution should be deep enough to cover at least one half of the hair coil. If it isn't, add more solution.

⚠ CAUTION
The basic and acidic solutions in this investigation are caustic. Be careful to rinse your hands after each step.

EXTRA
Vinegar is a dilute solution of acetic acid (CH₃COOH), usually between 3 and 5%.

EXTRA
The hair may be weakened and break easily.

Removing Color from Hair

6. Place 110 mL of bleach and 110 mL of water in a 400 mL beaker. Wrap 6 strands of the dark hair around a stirring rod in a spiral fashion. Tape the ends. Save the other 6 strands to use as a control for comparison purposes.

7. Submerge the hair coil in the diluted bleach solution. Allow it to stand for 10 minutes.

8. Remove the hair from the bleach and rinse under tap water for 3 minutes. Allow the hair to dry. Note the color change as you compare it to the control strands. Are there any changes in texture or strength?

Coloring Hair

9. Using 4 strands of the light-colored hair, place the hair on a watch glass. Drop blue food coloring along the length of the hair. Allow it to sit for 10 minutes while you proceed with the next step.

10. Submerge another 4 strands of the light hair in the bleach solution for 2 minutes.

11. Remove the strands from the bleach, thoroughly rinse with tap water, and place them on a watch glass. Drop blue food coloring onto the hair strands. Allow it to sit for 10 minutes and then drop some of the vinegar onto the hair.

12. After the hair in step 9 has been soaking in the food color for 5 minutes, remove it and compare the color to the remaining 4 strands of the light hair.

13. Rinse the hair under tap water for 1 minute. Note any color changes.

14. Wash the hair with a drop of shampoo. Note any color changes.

15. Repeat steps 12 through 14 for the hair treated in the base and then colored.

16. The solutions can be flushed down the drain with lots of water. The ammonia and the vinegar can be poured together, but the bleach should be flushed by itself.

EXTRA
Bleach is 5.25% sodium hypochlorite (NaClO).

EXTRA
If the hair is left too long, it could dissolve.

HAIR CHEMISTRY
PRE-LAB QUESTIONS

Name: _____

Lab Partner: _____

Section: _____   Date: _____

1) Why is hair without any residue preferred for the investigation?

2) The natural curl in hair is a function of the cross-sectional shape of hair. Rounder fibers are straight and the more oblong the fibers are, the tighter the curl. What would you expect to find if you viewed a microscopic section of a hair from a horse's tail? Hair from a poodle?

3) In what layer of the hair do you suppose this trait is expressed? The medulla, cortex, or cuticle? Support your answer.

4) People who spend a lot of time in swimming pools tend to have problems with the texture, strength, and color of their hair. To what would you attribute this?

5) Some hair types will not take a perm well. What might be the reason? (Hint: It can be a genetic thing.)

HAIR CHEMISTRY
REPORT SHEET

Name: _____

Lab Partner: _____

Section: _____   Date: _____

QUESTIONS

1. Examine the hair that was given the "permanent" and compare it to the control strands. Does the treated sample show more body and a greater tendency to curl on its own? Comment on the appearance, texture, and strength of the treated hair compared to the control sample.

2. Examine the hair that was bleached. Is the color lighter by 10%, 25%, 50%? (Estimate.)

3. How was the texture and strength of the hair changed?

4. Describe the ability of the normal strand to hold the color when rinsed in water and then when washed with shampoo.

5. Describe the ability of the hair that was treated with the base to hold the color when rinsed with water and then when shampooed. Explain what caused any differences you observe in the ability of the hair from the two coloring processes to hold the color.

6. Is the texture or strength of the treated and colored hair altered?

466

57 *Testing Sunscreens*

Light Can Be a Reactant

OBJECTIVES

To observe the photographic effect of light on a silver salt.

To obtain qualitative results from the reaction of sunscreens with ultraviolet light.

To determine which of the sunscreens tested provides the greatest protection from ultraviolet light.

Relates to Chapters 15 and 21 of Chemistry for Changing Times, *twelfth ed.*

BACKGROUND

Light can cause chemical reactions. Photosynthesis occurs when visible light interacts with chlorophyll. Vitamin D is produced in the skin of humans in a reaction induced by sunlight. Light also causes an electron transfer in photographic film, changing silver ions to black-colored silver atoms:

$$Ag^+ + e^- \rightarrow Ag(s) \text{ (black)}$$

A suntan can result when sunlight interacts with chemicals in the body to produce pigments that darken the skin color for protection. Too much sunlight can be damaging. A painful sunburn may occur if a person is in the sun too long without enough protection from either a suntan from previous exposure or applied chemicals. Some damage may occur even with the use of protective chemicals.

Sunscreens are chemicals that absorb sunlight and reemit the ultraviolet rays that are harmful to human skin at a different, less harmful, wavelength. They are rated with an SPF number, or skin protection factor. The number is the length of time required for protected skin to receive the same amount of sunlight as one hour without protection. What is sometimes forgotten by sun bathers is that the sunscreen may be washed off in water or by sweat. Many sunscreens need to be reapplied every 90 minutes.

Some sunglasses will block certain ultraviolet rays. Others are just darkened glass to reduce visible light and do not protect the eyes from the damaging rays.

In this investigation, a light-sensitive reaction will be used to test various sunscreens and sunglasses. The reaction occurs when the iron 3+ ion in the ferrioxalate ion reacts with light to produce an iron 2+ ion. The iron 2+ ion then reacts with the ferricyanide ion to produce a blue compound:

$$Fe(CN)_6^{3-} + Fe^{2+} \rightarrow Fe_3[Fe(CN)_6]_2 \text{ (blue)}$$

In this investigation, the silver compound is made by precipitation of silver chloride. When silver ions and chloride ions are together in solution, they will react to form an insoluble solid, called a *precipitate*.

Ultraviolet light is more energetic than visible light and will cause different chemical reactions to occur.

WASTE AND THE ENVIRONMENT

The compounds used in this investigation are not toxic. Care must be taken to keep metal ions out of the water system. Thus, metal ions should go to the toxic-waste site. In this case, the silver and iron compounds are not water-soluble, so they would not be dissolved if a landfill developed a leak.

⚠ *CAUTION*s warn about safety hazards.
*EXTRA*s give helpful hints, additional information, or interesting facts.

Reagents
   0.1 M hydrochloric acid [HCl]
   0.1 M silver nitrate [$AgNO_3$]
   potassium ferrioxalate [$K_3Fe(C_2O_4)_3 \cdot 3H_2O$]
   0.3 M potassium ferricyanide [$K_3Fe(CN)_6$]
   6 M nitric acid [HNO_3]
   1 M sodium hydroxide [NaOH]
   copper strip
   distilled water
   pH test paper

Common Materials
   white paper
   sunscreens, 3 different SPF factors
   sunglasses
   key (or other small solid object)
   paper towels

Laboratory Equipment
   UV light
   Petri dishes
   filter paper
   grease pencil
   eye droppers
   small test tube
   stopper to fit test tube

PROCEDURE

1. Place 3 drops of 0.1 M silver nitrate ($AgNO_3$) in the center of a piece of filter paper. Add 3 drops of 0.1 M hydrochloric acid (HCl) in the same place. Observe the resulting white solid.

2. Place a key or other object so that only a part of the white solid is covered. Place the filter paper under ultraviolet light for 5 minutes. The white solid should be dark only where light could react with the silver ion.

3. Add a few crystals of potassium ferrioxalate ($K_3Fe(C_2O_4)_3 \cdot 3H_2O$) to 2 mL of distilled water in a small test tube. Stopper the test tube and shake for 2 minutes. Add 2 mL of 0.3-M potassium ferricyanide [$K_3Fe(CN)_6$], stopper the test tube, and shake for a few seconds. Use the stopper to spread the mixture evenly over a white sheet of paper. Let the excess run off the paper. Using a paper towel, blot the white paper. Set the paper aside to dry.

4. Use a grease pencil on the outside of the bottom of a petri dish to draw a "+" sign, dividing the dish into four equal parts. Label the sections with the SPF numbers of three sunscreens that you will test. On the inside of the dish, coat each section with the corresponding sunscreen, leaving the fourth section uncoated. This section will serve as the control.

5. When the white paper from step 3 is dry, look at it under an ultraviolet light.

6. Place the petri dish on the white paper under the ultraviolet light. Place a pair of sunglasses on the white paper. Allow the paper to sit under the ultraviolet light for 5 minutes, then remove the glasses and the petri dish and complete the report sheet.

7. Place the used white paper in the trash. Flush all the other compounds down the drain with plenty of water. Place the silver chloride solid in a container to be buried in a secure toxic-waste landfill. An alternative method for silver disposal is to reclaim it as metal. Dissolve the silver salt in 6-M nitric acid (HNO_3). Neutralize with 1-M sodium hydroxide (NaOH) to pH 7. Place a clean copper strip in the solution. Set it aside for 20 minutes (or overnight). The silver metal will wipe off of the copper easily. Wipe the silver into a filter cone (see Investigation 9). Filter the silver metal. Allow it to dry.

⚠ _CAUTION_
Hydrochloric acid is corrosive.

EXTRA
A control is used to compare results.

EXTRA
Recycling is better than discarding.

TESTING SUNSCREENS
PRE-LAB QUESTIONS

Name: _____

Lab Partner: _____

Section: _____   Date: _____

1) There are areas of the globe where there are months with little or no daylight. What common deficiency in humans might you expect to find in those areas?

2) Why would it be advantageous to have your eyes protected from UV light rather than simply shielding them from the brightness of the light?

3) Can you think of other uses for the compound that reacts with UV light to turn blue?

4) Would an SPF rating mean the same thing on the equator as it would in New York City? Explain your reasoning.

TESTING SUNSCREENS
REPORT SHEET

Name: _____

Lab Partner: _____

Section: _____   Date: _____

I. SILVER
   1. Did all the white solid turn black? Why or why not?

   2. Photographic films are made with AgBr in a gel. What chemical reaction occurs when light interacts with the film?

II. SUNSCREEN TESTING
   Brand name   SPF rating

   _____   _____

   _____   _____

   _____   _____

III. QUESTIONS
   1. Which sunscreen provided the best protection from the UV rays?

   2. Which sunscreen provided the worst protection from the UV rays?

   3. Explain the visual appearance of the white paper when it was exposed to UV light.

4. How could further reaction of the chemicals of the white paper be stopped; that is, how can you "fix" the results?

5. Did the sunglasses block part of the UV rays?

58 *Lead Detection*

Get the Lead Out!

OBJECTIVES

To understand the toxic nature of lead compounds in the environment and some of the most
common sources.

To learn to calculate lead concentration in parts per million.

To detect the presence of lead in several materials.

To demonstrate the effect of lead on various proteins.

To observe the effect of lead on the action of an enzyme in saliva.

Relates to Chapters 13 and 22 of Chemistry for Changing Times, *twelfth ed.*

BACKGROUND

Lead is a dull gray metallic element that has an atomic number of 82 and an atomic mass of
207, making it one of the heavier elements with stable isotopes. Indeed, the naturally occurring
radioactive isotopes of thorium and uranium decay through a series of isotopes to *stable isotopes*
of lead. These stable isotopes are unaltered with time and remain indefinitely. Lead is very
dense, 11.3 g/cm^3, making it useful for ballast and for shielding materials from radiation.

Soft and malleable, lead is easy to shape; in fact it is so soft that it was the original writing
implement of choice, thus the name lead pencil, and was used until the sixteenth century when
graphite deposits were discovered in England. Graphite makes a darker mark than lead but is
brittle and therefore requires a rigid container—the wood of the pencil that we are accustomed to
using. The Romans first used lead for water pipes. Its Latin name, *plumbum*, is the origin of the
symbol Pb and is used for the name of people who work on water pipes: plumbers. Although the
softness of lead prevents its use in tools, it is desirable for use in bullets and low-melting alloys
to seal metal joints in water pipes and other plumbing needs, or to make electrical contact. Lead
and its compounds are major components of the automobile battery; lead weights are still used to
balance tires on automobiles; and lead compounds were added to some gasoline, referred to as
leaded gas, as an anti-knock additive. Tetraethyl lead was especially important for this purpose.
At one time, many paints contained a white pigment made of lead, and red lead was used as a
primer in the first coat over steel. Lead compounds are often used for the coloring of pottery
glazes, hence some pottery dishes, especially from foreign countries, may contain lead and
should not be used for food. Imported plastics and mini-blinds may contain lead that will slowly
leach out of the material.

Lead is not only stable but also fairly unreactive and therefore remains in the environment
for a very long time. These properties also make it quite difficult to clean up. Lead levels in soil
as small as 500 ppm (parts per million) classify the soil as "hazardous waste." *Parts per million*
is a mass ratio used to express small concentrations. For example, 500 ppm translates to about 1
lb of lead for every 1100 tons of soil.

The following uses of lead have declined because of its *toxicity*:
1. Water pipes are no longer made from lead, but lead-based solder is still a commonly
   used material for pipe joints.
2. The 94-year history of leaded paint ended in 1978.
3. The end of 57 years of production of leaded gasoline occurred in 1986.

The toxic effects of lead in the human body develop gradually, usually after repeated exposure over a period of time. The symptoms are not the same in everyone but often include a pale face and lips, a black line in the gums, mood swings, irritability, and severe abdominal cramps. In children, the brain is often involved, causing headaches, vomiting, irritability, convulsions, and coma. Lead poisoning can be fatal and when not fatal may cause attention deficits, learning disabilities, and mental retardation, especially in children.

One of the ways in which lead causes damage is by *denaturing*, or changing the shape of, a protein. Even a tiny amount of lead causes a response in a protein. The protein may no longer be able to perform its normal function. This is especially true of proteins that function as *enzymes* or biological *catalysts*.

This experiment will teach you how to test for lead samples that are at least 1% lead. Possible samples could be dust from the roadside, pencil paints, special-purpose lead paints, older paint chips, or an extract from glazed pottery. The reaction used involves the formation of a yellow precipitate according to the following reaction:

$$Pb^{2+} + 2\,I^- \rightarrow PbI_2\,(s).$$

Kits to detect lead in lesser concentrations are available from Hach Chemical Co., Frey Scientific Co., and others. These can be used to test fish, water, air, urine, food, and other samples.

The second part of the experiment will demonstrate the effect of lead on proteins. An enzyme in saliva breaks down starch into simple sugars. Iodine produces a blue color with starch but not with sugar. As long as the enzyme is active, starch is changed to sugar, and the blue color does not develop. When lead denatures the enzyme, starch is not converted to the simple sugar and so the blue color develops.

WASTE AND THE ENVIRONMENT

Lead, of course, is toxic and must be kept out of the water system. After the lead is removed, the other solutions are not toxic. Concentrated acidic or basic solutions can damage plumbing if not neutralized or diluted.

⚠ *CAUTION*s warn about safety hazards.

*EXTRA*s give helpful hints, additional information, or interesting facts.

Reagents

6 M nitric acid [HNO_3]

1 to 6 M acetic acid [CH_3COOH]

1% sodium sulfite [Na_2SO_3]

0.1 M potassium iodide [KI]

lead acetate [$Pb(C_2H_3O_2)_2$]

pH 7 buffer solution

pH test paper

1% starch solution

iodine solution

1% glucose solution

sodium sulfide [Na_2S]

1 M ferric chloride [$FeCl_3$] iron(III) chloride

sodium carbonate [Na_2CO]

Common Materials

egg

samples to test

Laboratory Equipment

1000 mL volumetric flask

50 mL beaker

centrifuge

hot water bath

10 mL graduated cylinder

filter paper

fume hood

large test tubes

balance

eyedropper

PROCEDURE
Part I: Detecting Lead

1. One pair of students should make up a lead solution to be used by the entire lab by weighing about 0.05 g of lead(II) acetate [$Pb(C_2H_3O_2)_2$] and dissolving it in 1 L of distilled water. Mark the container "lead(II) acetate solution." Each pair of students should take some lead solution to their desk in a small beaker.

2. Use a clean, dry 10 mL graduated cylinder to measure volumes. To 2 mL of the "lead(II) acetate solution," add a few drops of 0.1-M potassium iodide. The yellow precipitate is a positive test for lead.

3. Place 0.5 mL of 6 M nitric acid (HNO_3) in a test tube. Add 0.1 g of the sample to be tested. Place the test tube in a beaker of boiling water in a hood for 15 to 20 minutes in order to allow the lead sample to dissolve.

    ⚠ CAUTION
    Nitric acid is caustic.

4. To test for lead in pottery glazes, add 10 to 20 mL of acetic acid (CH_3COOH) to the pottery. Allow the pottery to stand for 1 hour. Transfer the acetic acid solution from the pottery into a test tube. Digest by placing the test tube with the sample in a beaker of boiling water in a hood for 5 to 10 minutes, as instructed in step 3.

    ⚠ CAUTION
    Toxic nitrogen oxide fumes are given off and should not be inhaled.
    Don't burn yourself.

5. Remove the test tube and sample used in step 1 or 2 from the hot water bath. Add 7 to 8 mL of freshly prepared 1% sodium sulfite (Na_2SO_3) to neutralize any excess nitric acid.

    ⚠ CAUTION
    Sulfur dioxide (SO_2) is given off and should not be inhaled.

6. Cool the sample. Centrifuge or let the sample stand until the precipitate has settled.

7. Add a few drops of 0.1 M potassium iodide (KI). A yellow precipitate indicates at least 1% lead in the sample.

Part II: Effect of Lead on Protein

8. Separate enough egg white from the yolk to make a quarter-size puddle on a watch glass. Separate the egg white by pouring the yolk from 1 shell half to the other shell half as the white drips into a beaker or a watch glass.

    EXTRA
    Just as you would when separating egg whites in the kitchen.

9. Using an eyedropper, drop about 1 mL (20 drops) of lead(II) acetate solution onto the clear egg white. Egg white contains a protein called *albumin*. Observe and record the reaction.

Part III: Effect of Lead on an Enzyme

10. Each person should collect about 1 mL of saliva in a clean test tube. Pour the saliva from both partners into a small beaker. Rinse both test tubes into the beaker with about 10 mL of pH 7 buffer solution. Add enough distilled water to bring the volume to 30 mL of solution in the beaker. Stir until all is dissolved.

EXTRA
Think of your favorite foods to stimulate saliva production.

11. Pour 10 mL of the saliva solution into each of 3 large test tubes.

12. To test tube 1, add 10 mL of 1% starch solution and 5 mL of iodine solution. Note the color. Set the test tube aside but continue to monitor the color.

EXTRA
An enzyme in saliva breaks down starch into simpler sugars. The starch reacts with iodine to make a blue color. Simple sugars do not produce the blue color.

13. To test tube 2, add 1 mL of lead solution. Allow the lead and saliva to react for 15 seconds, then add 10 mL of 1% starch solution and 5 mL of iodine solution. Note the color. Set the test tube aside but continue to monitor the color.

14. To test tube 3, add 10 mL of 1% glucose solution and 5 mL of iodine solution. Note the lack of color

15. Collect all used solutions containing lead. Dissolve the lead iodide (PbI_2) in 6 M nitric acid. Add a threefold excess of sodium sulfide (Na_2S). Let the solution sit for an hour. Adjust the pH of the solution to neutral (pH 7 or more). Use pH paper to check the pH. Filter the precipitate using a filter paper prepared according to the diagram below. Discard the precipitate and filter paper in a container to be buried in a secure toxic-waste landfill. The filtrate contains sulfide. Remove the sulfide by adding 1-M iron (III) chloride (ferric chloride) in a threefold excess. Neutralize the solution with sodium carbonate (Na_2CO_3). Filter the precipitate. Put the precipitate and filter paper in a container to be buried in a toxic-waste landfill. Flush the remaining filtrate and the other solutions down the drain with plenty of water.

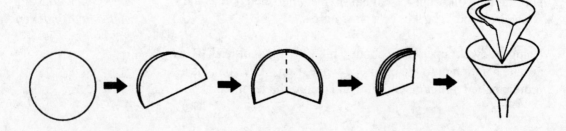

LEAD DETECTION
PRE-LAB QUESTIONS

Name: _____

Lab Partner: _____

Section: _____   Date: _____

1) Lead is commonly used in "shot" to load shotgun shells for such sports as quail hunting. Evaluate this practice in light of environmental concerns. Suggest an inexpensive alternative.

2) Why would soils along highways still contain a high level of lead even though it has not been added to gasoline since 1986?

3) Recent news stories have revealed a high level of lead in some trinkets and inexpensive jewelry imported from other parts of the world. These items are often found in toy-dispensing machines in grocery and variety stores. Is there a potential risk to children from these items?

4) What is the relationship between pencil lead and lead metal?

LEAD DETECTION
REPORT SHEET

Name: _____

Lab Partner: _____

Section: _____   Date: _____

I. DETECTING LEAD

1. Sample: _____

    Positive lead test   Yes _____   No _____

2. Lead Concentration

$$\underline{\hspace{2cm}} \text{ g Pb}(C_2H_3O_2)_2 \times \frac{207 \text{ g PB}}{325 \text{ g Pb}(C_2H_3O_2)_2} = \underline{\hspace{2cm}} \text{ g Pb}$$

$$\frac{\underline{\hspace{1.5cm}} \text{ g Pb}}{1000 \text{ mL}} \times \frac{1 \text{ mL solution}}{1 \text{ g solution}} \times 10^6 = \underline{\hspace{2cm}} \text{ ppm Pb}$$

II. EFFECT ON A PROTEIN
1. Describe the effect of lead on albumin.

2. One emergency treatment for ingesting lead is to swallow egg white or milk. Why?

3. Why is administration of an emetic (vomit inducer) the next step of the treatment?

481

Chemical Investigations for Changing Times

III. EFFECT ON AN ENZYME
1. Describe the color change that occurred in each of the three test tubes.

2. Did the amylase in the saliva sample break down the starch in test tube 2? Explain.

IV. QUESTIONS
1. What is the purpose of the digestion step in step 2?

2. The lead content of drinking water is a concern in many areas of the U.S. List possible sources of lead contamination of our water supplies.

3. What was the purpose of step 14 in the enzyme experiment?

482

59 *Parts per Million*

How Much Is Really There?

OBJECTIVES

To gain an appreciation for the need to measure minute concentrations of substances.

To relate the measurements of LD_{50} and LC_{50} to toxicity and concentration.

To understand the concentration expressions of parts per million and parts per billion.

To prepare a very dilute solution of a copper(II) chloride and then dilute it to ppm and ppb concentrations.

To determine the amount of chloride ion in several solutions by precipitation with a silver ion.

Relates to Chapter 22 of Chemistry for Changing Times, *twelfth ed.*

BACKGROUND

Many compounds and elements are known to be poisons. Other materials are toxic in high concentrations but have no effect at a low concentration. When a compound is found to be toxic, many people suggest the "allowable concentration should be zero." Scientists continue to develop methods of determining smaller and smaller concentrations. As a result we know that there are no perfectly pure substances. That is, there are always trace amounts of random particles in any material. Our abilities to detect extremely small concentrations of impurities have made us aware of the inevitable presence of many things in our environment, such as in our drinking water. For example, it is assumed that any toxic heavy metals in the environment will eventually find their way into ground water. The question has now become, "What is an acceptable concentration?"

Concentrations are often measured in terms of molarity, that is, moles of solute per liter of solution. Small concentrations are often measured in units of either parts per million (ppm) or parts per billion (ppb). A *part per million* is one gram of a particular material in one million grams of total material. For instance, one person living in a city of a million is one part per million.

The toxicity of many compounds have been determined. Often that value is given in terms of LD_{50}, which is the amount of material that would be a *lethal dose* to 50 percent of the population. LD_{50} values are used for solid and liquid substances and the units are often in milligrams per kilogram of body mass, mg/kg. Some lethal doses are large (more than a gram), but some compounds have lethal doses that are in the nanogram range. Another measurement is LC_{50}. It is used to measure *concentrations* of substances in the air (and sometimes in water) that are lethal to 50% of a population after an extended time of exposure, usually four hours. Typical units for the LC_{50} measurement are ppm. A solution concentration of 1 ppm is the same as 1 mg of solute per kilogram of solution. As an example, consider tetraethyl lead, a substance added to gasoline prior to 1986. Lead still remains in the environment. The LD_{50} value for tetraethyl lead ingestion in rats is roughly 1.2 mg/kg and the LC_{50} value for inhalation by rats is 850 mg/m^3— or, since air has a density of 1.29 kg/m^3, 660 ppm.

We can illustrate the effect of concentration on a reaction using the chloride ion. Certain concentrations of chloride ion can be detected by a precipitation reaction with the silver ion as follows:

$$Ag^+ + Cl^- \rightarrow AgCl\ (s).$$

However, if the concentration of chloride is too low, this reaction will not occur, no solid, called a precipitate, will form.

We can sometimes use the detection of one ion concentration, in a solution containing a specific ratio of ions, to determine indirectly the amount of another ion in the solution. In this investigation, we will use a test for the presence of one ion in a solution to infer the presence of another.

WASTE AND THE ENVIRONMENT

The compounds used in this investigation are not toxic; however, it is not good practice to place certain metal ions into the water system. Your instructor will provide instructions for disposal of solutions.

▲ *CAUTION*s warn about safety hazards.

*EXTRA*s give helpful hints, additional information, or interesting facts.

Reagents

copper(II) chloride [$CuCl_2$]   1 M sodium hydroxide [NaOH]
1-M silver nitrate [$AgNO_3$]   copper strip
6 M nitric acid [HNO_3]   pH test paper
distilled water

Laboratory Equipment

1000 mL volumetric flask   filter paper
10 mL pipet with bulb   balance
50 mL beaker   dropper
10 mL graduated cylinder   3 1000 mL beakers

PROCEDURE

1. Weigh a clean, dry 50 mL beaker. Add about 0.014 g of copper(II) chloride ($CuCl_2$). Weigh the beaker and copper(II) chloride.

2. Dissolve the copper(II) chloride in distilled water. Pour the solution into a 1000 mL volumetric flask. Rinse the beaker into the flask several times with distilled water. Fill the flask to just below the etched line. Invert the flask several times to mix the solution. Add 1 drop of water at a time until the bottom of the meniscus (the curved top of the liquid) is even with the etched line.

   EXTRA
   The blue color is due to the copper(II) ion.

3. Pour this solution into a large beaker marked "1." Rinse out the volumetric flask with distilled water several times, discarding it in the sink each time.

4. Using a 10 mL graduated cylinder, transfer exactly 10 mL of solution 1 to the volumetric flask. Fill the flask with distilled water as before.

5. Pour this solution into a second large beaker marked "2." Rinse out the volumetric flask and graduated cylinder with distilled water several times.

6. Using the 10 mL graduated cylinder, transfer exactly 10 mL of solution 2 into the volumetric flask. Fill the flask with distilled water as before. Pour this solution into a large beaker marked "3."

7. Pour about 10 mL of solution 1 into a test tube. Add, a drop at a time, 1-M silver nitrate ($AgNO_3$) solution. Count the number of drops necessary for a precipitate to appear. Record the number of drops.

8. Repeat step 7 using solution 2 and then solution 3. If you reach 50 drops and no precipitate has appeared, stop and record that information.

9. Filter the solutions and discard the solids and filter paper in a silver waste container. The copper solutions should be evaporated and the solid buried in a secure landfill. However, in this investigation, the concentrations are so small they are not a threat to the environment. An alternative method for silver disposal is to reclaim it as metal. Dissolve the silver salt in 6-M nitric acid (HNO_3). Neutralize the solution with 1-M sodium hydroxide (NaOH) to pH 7. Place a clean copper strip in the solution. Allow the solution to sit for 20 minutes (or overnight). The silver metal will wipe off the copper easily. Wipe the silver into a filter cone (see step 2 of Investigation 8). Filter the silver metal and allow it to dry.

EXTRA
Recycling is better than discarding.

PARTS PER MILLION
PRE-LAB QUESTIONS

Name: _____

Lab Partner: _____

Section: _____   Date: _____

1) Lead was an additive to gasoline for over half a century and is now found in the environment in relatively high concentrations, especially along roadways. Why is that a concern?

2) Explain why it is unreasonable to expect that there will be absolutely no toxins in anything we take from the environment.

3) Discuss the importance of determining toxicity levels of various substances for all animals in the food chain, including humans.

4) Why is it important that scientists be able to detect minute concentrations of substances in our air and water?

PARTS PER MILLION
REPORT SHEET

Name: _____

Lab Partner: _____

Section: _____   Date: _____

I.   Mass of $CuCl_2$ + beaker   _____g

Minus mass of the beaker   –   _____g

Mass of $CuCl_2$ _____g

Solution 1

Molarity: $$\frac{g\,CuCl_2}{1000\,mL\,solution} \times \frac{mL}{10^{-3}\,L} \times \frac{mol}{135\,g}$$ = _____ M

Assume that the solution density is 1.00 g/mL.

ppm: $$\frac{g\,CuCl_2}{1000\,mL\,solution} \times \frac{mL\,solution}{1.00\,g\,solution} \times 10^6$$ = _____ ppm

Color of solution _____

# of drops for precipitation _____

Solution 2

Molarity: _____ M solution #1 $\times \dfrac{10\,mL}{1000\,mL}$ = _____ M

ppm: _____ ppm $\times \dfrac{10\,mL}{1000\,mL}$ = _____ ppm

Color of solution _____

# of drops for precipitation _____

Solution 3

Molarity: _____M solution #2 $\times \dfrac{10\ mL}{1000\ mL}$ = _____M

ppm: _____ppm $\times \dfrac{10\ mL}{1000\ mL}$ = _____ ppm

ppb: _____ppm x 1000 = _____ ppb

Color of solution _____

# of drops for precipitation _____

II. QUESTIONS

1. Where the chloride ion was detected, what can be said about the copper(II) ion in solution?

2. Could the chloride in solution 3 be detected by this method?

3. Were there copper(II) ions in solution 3?

4. Can the presence of copper(II) ions be detected by the color of the solution in 1?

   In the solution in 2?

   In the solution in 3?

   Explain.

5. If the copper(II) ions had a toxicity level similar to lead, would solution 3 be safe? Explain your reasoning.

6. Why was it important to use distilled water to rinse the labware in this investigation?

I Common Laboratory Apparatus*

Forceps are for handling hot items or small items that could be contaminated by touching (either chemically or by changing their weight).

Tongs are similar to forceps in function but useful for larger items.

Beakers are useful as reaction containers or to hold liquid or solid samples. They are also used to catch liquids from titrations and filtrates from filtering operations.

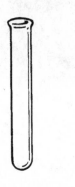

*Test tube*s are for holding small samples or for containing small-scale reactions.

Droppers are for adding liquids one drop at a time.

*Figures are not drawn to scale.

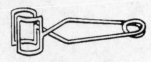

Test tube holders are for holding test tubes when the test tubes should not be touched.

Evaporating dishes are for holding hot samples or for drying samples.

Watch glasses are for holding small samples or for covering beakers or evaporating dishes.

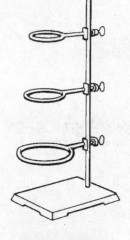

Ringstands with rings are for holding pieces of glassware in place.

Wire gauze is useful for placing on rings to hold beakers that are to be heated with a Bunsen burner.

Bunsen burners are sources of heat.

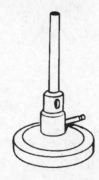

Clay triangles are placed on a ring on a ringstand as a support for a funnel.

Glass funnels are for transferring liquids from one container to another or for filtering when equipped with filter paper.

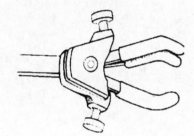

Clamp holders are for clamping various pieces of hardware to the ringstand.

Clamps are for securing various hardware to the ringstand.

Buchner funnels are for separating quantities of solids from a volume of liquid. The correct size of filter paper disk must be used. The funnel is held in the filter flask by a one-hole stopper.

Filter flasks are receptacles for filtrates from Buchner funnels. The sidearm can be connected to an aspirator via rubber tubing so suction can be applied to the system to increase the rate of filtration.

Burets are for adding a precise volume of liquid. The volume of liquid added can be determined to the nearest 0.01 mL with practice. Although common in many laboratories, this manual does not make use of a buret.

Graduated cylinders are for measuring an amount of liquid. The volume of liquid can be estimated to the nearest 0.1 mL with practice.

Erlenmeyer flasks are useful for containing reactions or for holding liquid samples. They are also useful for catching filtrates.

II Common Laboratory Techniques

1. Test-Tube Safety

When attempting to sense an odor, never inhale directly—that is, don't place your nose directly over the test tube. Waft the fumes toward you with your hand.

When heating a liquid in a test tube, gently heat the test tube at the top of the liquid, keeping the test tube moving. If you heat one spot, the liquid may turn to a gas and blow the entire contents out the top of the test tube. It is usually easier to heat the test tube by keeping it at an angle. To prevent spraying anyone, never point the test tube at yourself or others.

2. Acid Dilution

When an acid needs to be diluted, always pour the acid into water. Enough heat can evolve when an acid is diluted to cause drops of solution to jump out of the beaker. If acid is being poured into water, there is less chance of an acid burn. Pour the acid slowly to reduce the amount of heat evolved and the splattering.

3. Care of Reagents

The reagents used by the entire lab can be contaminated by one careless student. The best rule is never put anything into the reagent bottle. If you need a certain amount of liquid, either pour from the reagent bottle into a beaker and draw up the liquid into a pipet from the beaker or pour from the beaker into a graduated cylinder, adding the last drop with an eyedropper filled from the beaker. Never put a spatula into a reagent bottle. Pour the solid into a beaker or onto weighing paper. Never pour excess reagent back into the bottle.

4. Titration

A titration procedure is performed by adding small amounts of one reactant to a beaker or flask containing the other reactant until the reaction is completed. In an acid-base reaction, the reaction is complete when the amount of acid equals the amount of base. The experimenter is aware of equivalence by the change in pH that occurs. The change in pH is indicated by a color change of an indicator or by a pH meter. In other types of reactions, the completion of the reaction is usually indicated by a color change.

Titrations are usually performed using a buret. Because using a buret well requires some practice, in this manual the procedures are written to use a squeeze bottle instead of a buret. The solution to be used to titrate with is placed in the squeeze bottle after its density is determined. Weighing exactly 10 mL of solution is an easy way to determine the density. Weigh the filled squeeze bottle. To titrate, add solution from the squeeze bottle to the other solution in a flask. Often, a localized color change will be evident for a short period of time. Mixing the solution by swirling the flask will usually cause the color to return to the original color. As the color lasts longer make the next addition of solution smaller. Close to the endpoint (reaction completion), add the reagent one drop at a time. When the endpoint is reached, weigh the squeeze bottle. The volume added is the difference in weight of the bottle times the density of the solution.

5. Meniscus

In all volumetric glassware (pipet, buret, volumetric flasks, graduated cylinder), it is necessary to read the level of a liquid. A liquid in a small diameter container will form a meniscus or curve at the top surface. Usually this meniscus curves downward to a minimum at the center. To read the level of the liquid properly, your eye should be at the same level as the bottom of the meniscus. Sometimes a white card or a white card with a black mark on it will help you to see the meniscus clearly. For volumetric flasks and transfer pipets, the volume of the glassware is exact when the bottom of the meniscus is even with the etched line. In a graduated cylinder or a buret, the volume is read from the graduations etched on the glass. To read the volume correctly, visualize the distance between the tenths of milliliter marks as divided into 10 equal segments. Find the volume by reading the number of milliliters and number of tenths of milliliters and estimating hundredths of milliliters.

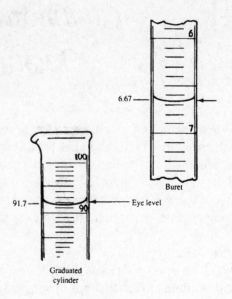

6. Care of Balance

Never weigh chemicals directly on the balance pan. Always use weighing paper, weighing boats, or glassware. The chemicals can corrode the balances, which will cause an error in weight. Never pour liquid into a beaker on the balance. It is too easy to spill onto the balance, even into the delicate mechanism. Creased weighing paper will contain the material more efficiently than flat paper. Always clean off the balance after you use it to prevent corrosion.

7. Bunsen Burner

The key to adjusting a Bunsen burner is to adjust simultaneously the gas from the gas jet and the air from the air vents on the side. For a clean-burning flame, more air is required than is available at the gas exhaust. The air vents and gas can be adjusted to produce an inner blue cone in the flame with no yellow tip. When cutting back on the gas supply also cut back on the air vents to prevent the flame from moving downward into the burner. Too little air produces a sooty, orange yellow-tipped flame. Too much gas may cause the flame to separate from the burner and even blow itself out.

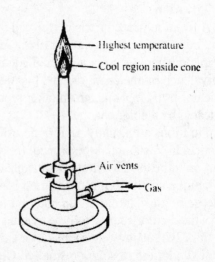